BENEATH A PETROLIFEROUS MOON

STANDARD OIL CO.

Pablo Neruda

When the drill bored down toward the stony fissures
and plunged its implacable intestine
into the subterranean estates,
and dead years, eyes of the ages,
imprisoned plants' roots
and scaly systems
became strata of water,
fire shot up through the tubes
transformed into cold liquid,
in the customs house of the heights,
issuing from its world of sinister depth,
it encountered a pale engineer
and a title deed.

However entangled the petroleum's arteries may be,
however the layers may change their silent site
and move their sovereignty amid the earth's bowels,
when the fountain gushes its paraffin foliage,
Standard Oil arrived beforehand
with its checks and its guns,
with its governments and its prisoners.

Their obese emperors from New York
are suave smiling assassins
who buy silk, nylon, cigars
petty tyrants and dictators.

They buy countries, people, seas, police, county councils,
distant regions where the poor hoard their corn
like misers their gold:
Standard Oil awakens them,

clothes them in uniforms, designates
which brother is the enemy.
the Paraguayan fights its war,
and the Bolivian wastes away
in the jungle with its machine gun.

A President assassinated for a drop of petroleum,
a million-acre mortgage,
a swift execution on a morning mortal with light, petrified,
a new prison camp for subversives,
in Patagonia, a betrayal, scattered shots
beneath a petroliferous moon,
a subtle change of ministers
in the capital, a whisper
like an oil tide,
and zap, you'll see
how Standard Oil's letters shine above the clouds,
above the seas, in your home,
illuminating their dominions.

— Pablo Neruda, *Canto General*, 1940
Translated by Jack Schmitt

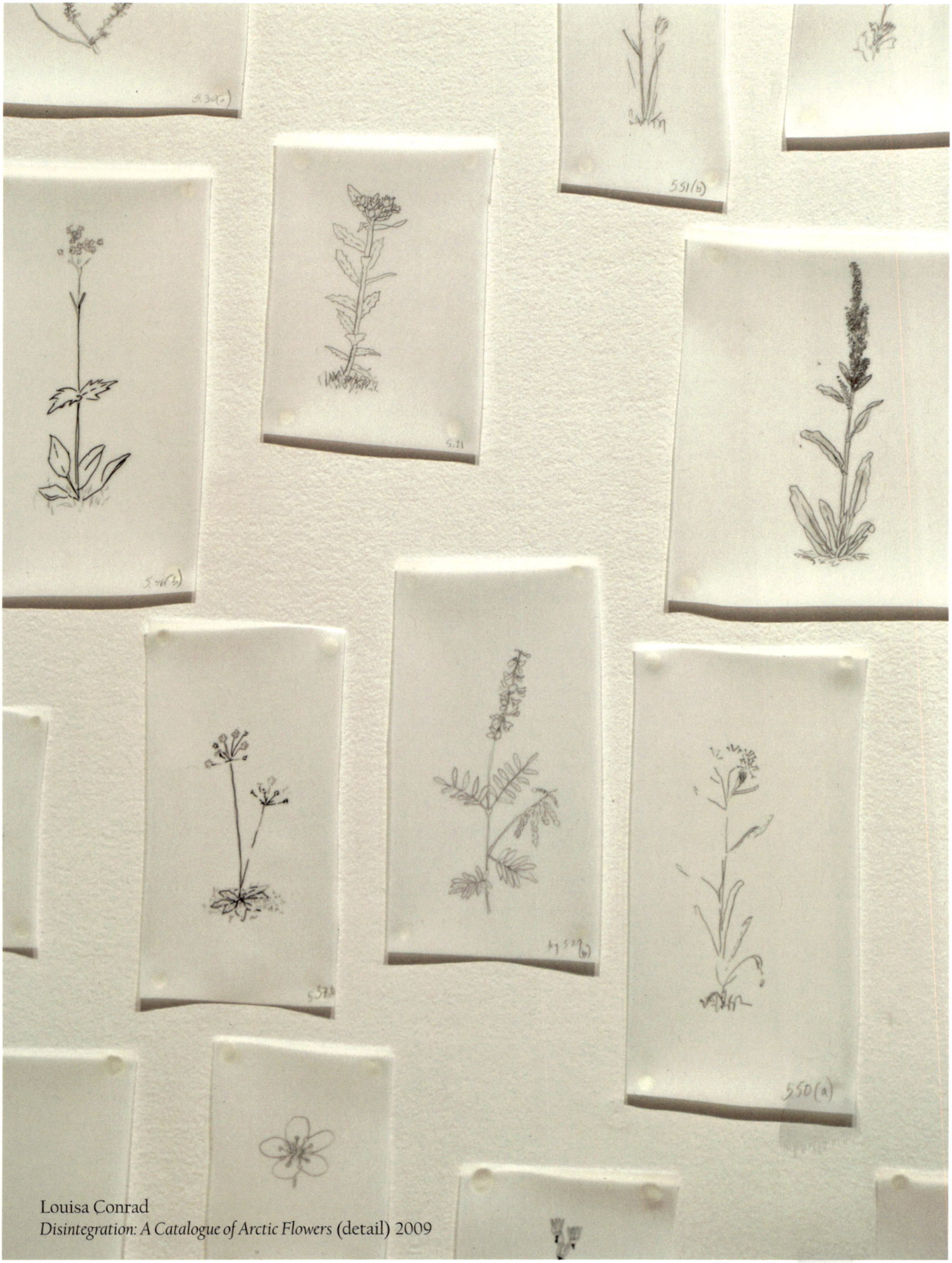

Louisa Conrad
Disintegration: A Catalogue of Arctic Flowers (detail) 2009

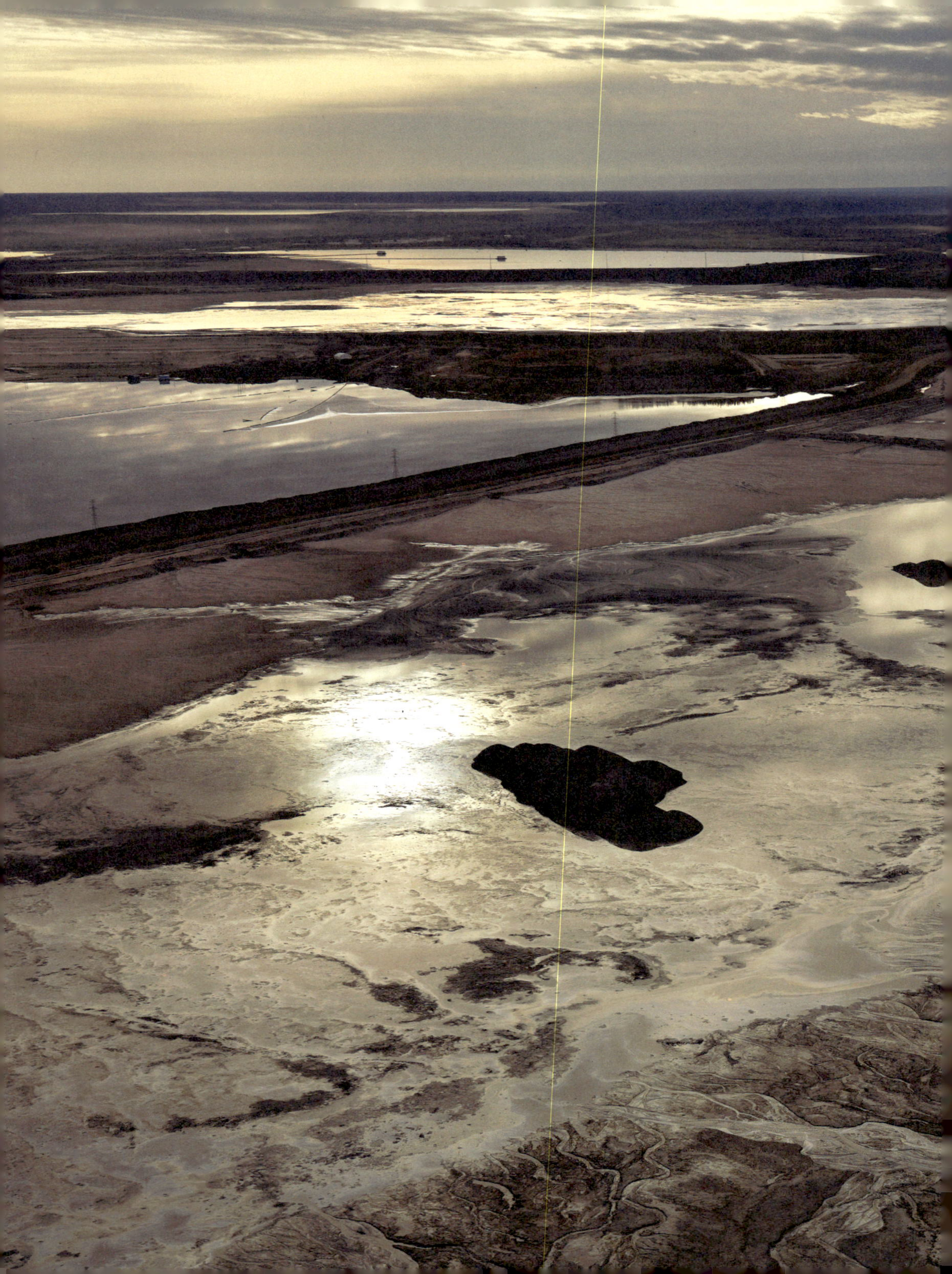

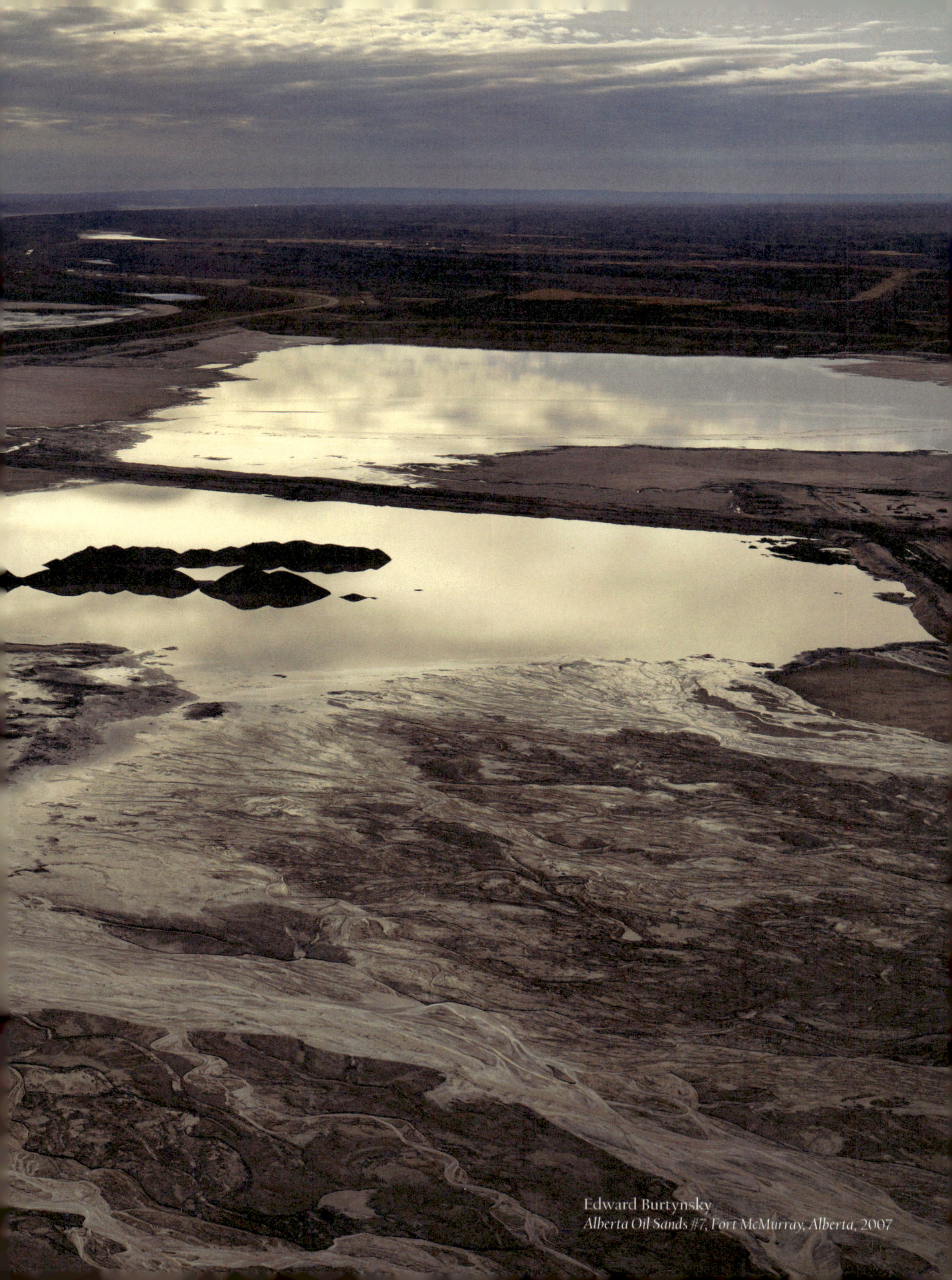

Edward Burtynsky
Alberta Oil Sands #7, Fort McMurray, Alberta, 2007

FOREWORD

SINCE 2006, SASKATCHEWAN has been enjoying a period of unprecedented economic growth, thanks largely to its abundant natural resources, including potash, uranium, and oil. While the province is well known for its production of the first two resources, most people are quite unaware that, amongst Canadian provinces, only Alberta exceeds Saskatchewan in overall oil production. Petroleum products extracted here include heavy and light crude, as well as natural gas; oil sands in the northern part of the province are also slated for development.

Beneath a Petroliferous Moon was initiated in response to the growing importance of the petroleum industry in Saskatchewan's economy. A line from a 1940 poem by Chilean poet Pablo Neruda served as the exhibition's title and set the tone for Associate Curator Jen Budney's global survey of artistic responses to the subject of 'big oil.' From across North America, Europe, and Africa, the art works in this exhibition embodied a variety of genres, from drawing to photography, and from research to the ready-made.

Realizing that dialogue can only emerge from the sharing of interests and concerns, *Beneath a Petroliferous Moon* explored the uncertainties of human reliance upon a particular class of fossil fuel at a moment when society's understanding of the negative impacts of its extraction, refining, and combustion have utterly surpassed our collective will and ability to implement alternatives. Many of the art works in the exhibition acknowledged an inescapable ambivalence towards an industry that produces such great wonders at such great cost. The exhibition, as a whole, attempted to open a space for visitors to reflect upon current issues in petroleum production and usage not just in relation to ecology, but also from cultural, historical, political, economic perspectives, all the while keeping beauty and a sense of awe in mind.

Beneath a Petroliferous Moon articulated the ambition of the Mendel Art Gallery to contribute locally, nationally, and internationally to the presentation, development, and discussion of contemporary art. The dedicated commitment of staff at the Mendel Art Gallery was essential to the success of the exhibition. It was ably and passionately assembled by Jen Budney, while its reception by a diverse array of audiences was made possible through the hard work and commitment of Public Programs Coordinator Laura Kinzel and her team.

The Mendel Art Gallery is grateful to all the institutions and individuals who loaned work for this exhibition. We are also grateful to the funding agencies and sponsors who support our work. Most importantly, I extend thanks to the eleven artists who exhibited their work in this exhibition, many of whom showed their work at the Mendel for the first time. A few of the artists came to Saskatoon for the installation and opening, adding to the rich dialogue between the Gallery's curatorial and public programming staff and members of the community. This catalogue, featuring an essay by the exhibition curator, along with an introduction by world-renown energy expert Vaclav Smil, and poems by Marcella Durand and the late Pablo Neruda, has been designed as a substantial record of the artists' contributions, all produced "beneath a petroliferous moon."

— Gregory Burke, Executive Director and CEO, Mendel Art Gallery

TWO SIDES
OF A
BARREL

Vaclav Smil

THIS IS NOT A LARGE EXHIBITION, BUT IT SHOWS plenty of oil's awful side: landscapes ravaged by the excavation and processing of Alberta's oil sands in the large colour photographs by Edward Burtynsky; dark effluents polluting waters in the drawings of Susan Turcot; grotesque sculptural evocations of drilling rigs made from oil-based plastics washed up on Scotland's beaches by Austria's Ernst Logar. There are 'masks' made of discarded plastic gasoline canisters and other waste from the West African oil industry by Yoruba artist Romuald Hazoumè. As a quart of oil spills over his camera lens, Robert Ladislas Derr's video focuses on a bank in Oil City, Pennsylvania, where the modern oil era began 150 years ago. And *The Last Arctic* photographs by Louisa Conrad remind us that the oil industry is poised to desecrate some of the planet's most remote, pristine areas of silent beauty. Yet the devastating effects of the oil industry — both human and environmental — are only part of a much more complex reality.

Like the generation of electricity, modern food production, or the growing web of wireless communication, the petroleum industry has two faces. One is worrisome enough to be portrayed with concern, condemnation, and even revulsion. When not done right, oil extraction despoils land and pollutes fresh and ocean waters. Oil refining is a dangerous enterprise as flammable liquids and gases under pressure can easily explode. Also, the combustion of liquid fuels is a major source of atmospheric air pollution and a leading contributor of carbon dioxide, the most significant anthropogenic greenhouse gas.

The other face of the oil industry shows that crude oil is the most important energy source in the modern world. In 2012 crude oil supplied a third of the world's commercial energy, ahead of coal (about 30 per cent) and natural gas (about 24 per cent), with the rest coming mostly from primary electricity (hydro, nuclear, solar, and wind). Our modern civilization was created and is maintained by large, incessant flows of fossil energies. Indeed, we are able to take the ubiquity, reliability, and indispensability of all of our services and benefits for granted precisely because we are sheltered from other realities by our reliance on oil.[1]

Exports and imports of oil are the world's most valuable trading activity, hence the foundation of many economies. No manufacturing activity has been as important in creating the modern industrial world as the production of gasoline- and diesel-fuelled vehicles. The mass ownership of cars helped to expand, or sometimes to create, entire economic sectors in affluent countries. Although railways can be completely electrified, intercontinental transportation via ships and planes is totally dependent on liquid fuels, and so are all heavy trucks and diesel-powered trains around the world. Liquid fuels have an even more important role in powering the tractors, combines, and irrigation pumps employed to grow food for the planet's seven billion people. Equally important, oil as a feedstock in chemical industries is transformed into new substances — above all, plastics and nitrogen fertilizers — which are thoroughly woven into the fabric of modern life.

In Canada, our relationship to oil is more important and more existential than in other affluent nations, and not just because we produce and export a great deal of it, or because our country is so cold (we do most of our heating with inexpensive natural gas or hydroelectricity, not with oil). Rather, it's partly attributable to the fact than 80 per cent of Canadians live in cities that are separated from each other by hundreds, even thousands, of kilometres of forests, fields, and mountains. The most convenient way to make connections between those cities is, by far, flying (for people) or by trucks (for commercial goods, especially foodstuffs). Furthermore, people living in isolated houses or small settlements far away from cities have only their trucks and cars to connect them with the rest of the country. This reality is particularly significant in rural Saskatchewan, where cultivated land goes further north and away from major cities than in Manitoba, Ontario, or Quebec.

Canadian annual per capita oil consumption is now more than 23 barrels. For easy reference, just look at the five barrels stacked in the centre of this exhibition: their size is correct; only their exteriors have been altered, colourfully painted by Jimmie Durham. Twenty-three barrels translates to about 2.8 tonnes or forty times the mass of a 70-kilogram person. Americans now consume about 2.6 tonnes a year per capita. Germans require only 1.35 tonnes per capita annually. This may seem surprising, considering the fuel they burn on their *Autobahnen*, which have no speed limit, but the country is densely inhabited, with 82 million people fitting into an area that is only a tiny bit larger than half of Saskatchewan (population one million).

Yet Sweden, a country that is, on average, about as cold as Canada and also relatively quite large (its latitudinal span is just over 1,500 kilometres, 20 per cent longer than Saskatchewan's south-north span), individuals consume less than 1.7 tonnes annually. So Canada's geographical size is not the only reason for our extraordinarily high per capita oil consumption. There are two other reasons. First, our rails: despite the fact that until recently, when we were surpassed by China, we have been the world's largest producer of hydroelectricity, Canada's transcontinental railroads are not electrified. We have no excuse for our failure to replace diesel locomotives with cleaner and more efficient electric traction energy. All the long-distance railroads in Europe are electrified, including the Trans-Siberian railroad in Russia, which is by far the longest in the world. Second, our travel habits: during the peculiar annual mass migration of Canada's population, millions of people drive and fly to Florida, Texas, the desert states of the United States Southwest, Mexico, smaller countries of Central America, and the islands of the Caribbean, consuming millions of tonnes of refined liquid fuels in the process.

Eventually, the world will move away from fossil fuels and begin to harness renewable energy flows, or a new, better version of nuclear electricity. But energy transitions are always gradual, protracted affairs, and for the next two or three generations (40 to 60 years), the world's dependence on oil will remain high, particularly in Canada. We should do our utmost to minimize the undesirable consequence of this dependence — but to think that we can free ourselves of it easily and speedily would be a delusion.

We are faced with a very large challenge to create a widespread, informed, and balanced understanding of modern civilization's dependence on fossil fuels, of their numerous benefits and their worrisome impacts. Most people will not acquire this understanding by applying themselves to the study of energy engineering, energy use, or the environmental impacts of fuels. That is why an art exhibition that displays captivating photographs, sculptures, and installations about petroleum is an excellent way to engender more than an ephemeral interest. The exhibition *Beneath a Petroliferous Moon* poses some intriguing questions that might initiate a learning process, which will lead the people who see these works to a better understanding of the complexities of modern oil addiction.

NOTES

1. Because of our abundant natural gas resource, the consumption share of crude in Canada is slightly lower (just over 31 per cent), but we are also a major exporter of the fuel, selling about half of what we produce

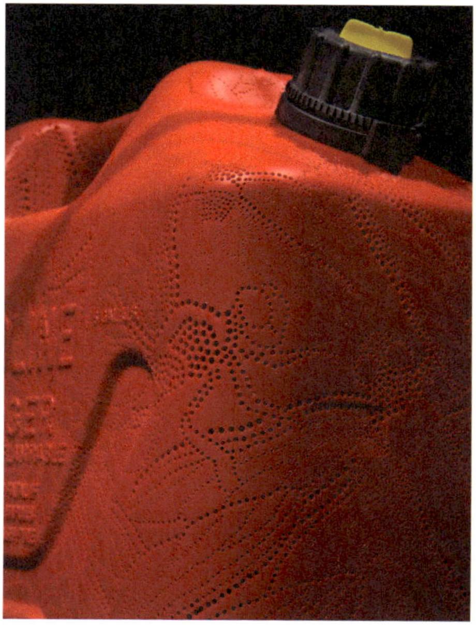

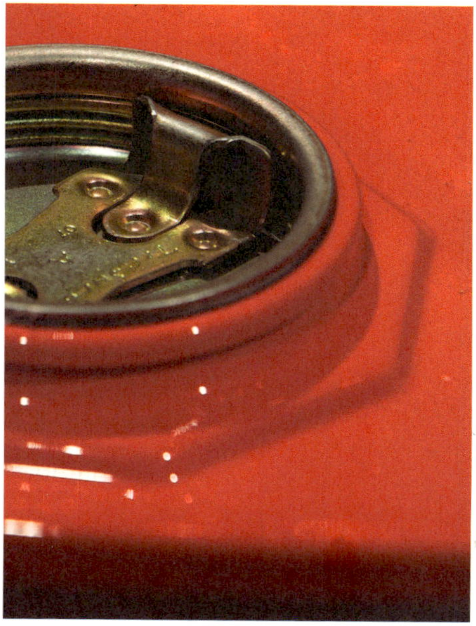

Brian Jungen, *Dragonfly* (detail) 2008, and Jimmie Durham, *Sweet light crude (Dubai–Oman)* (detail), 2008

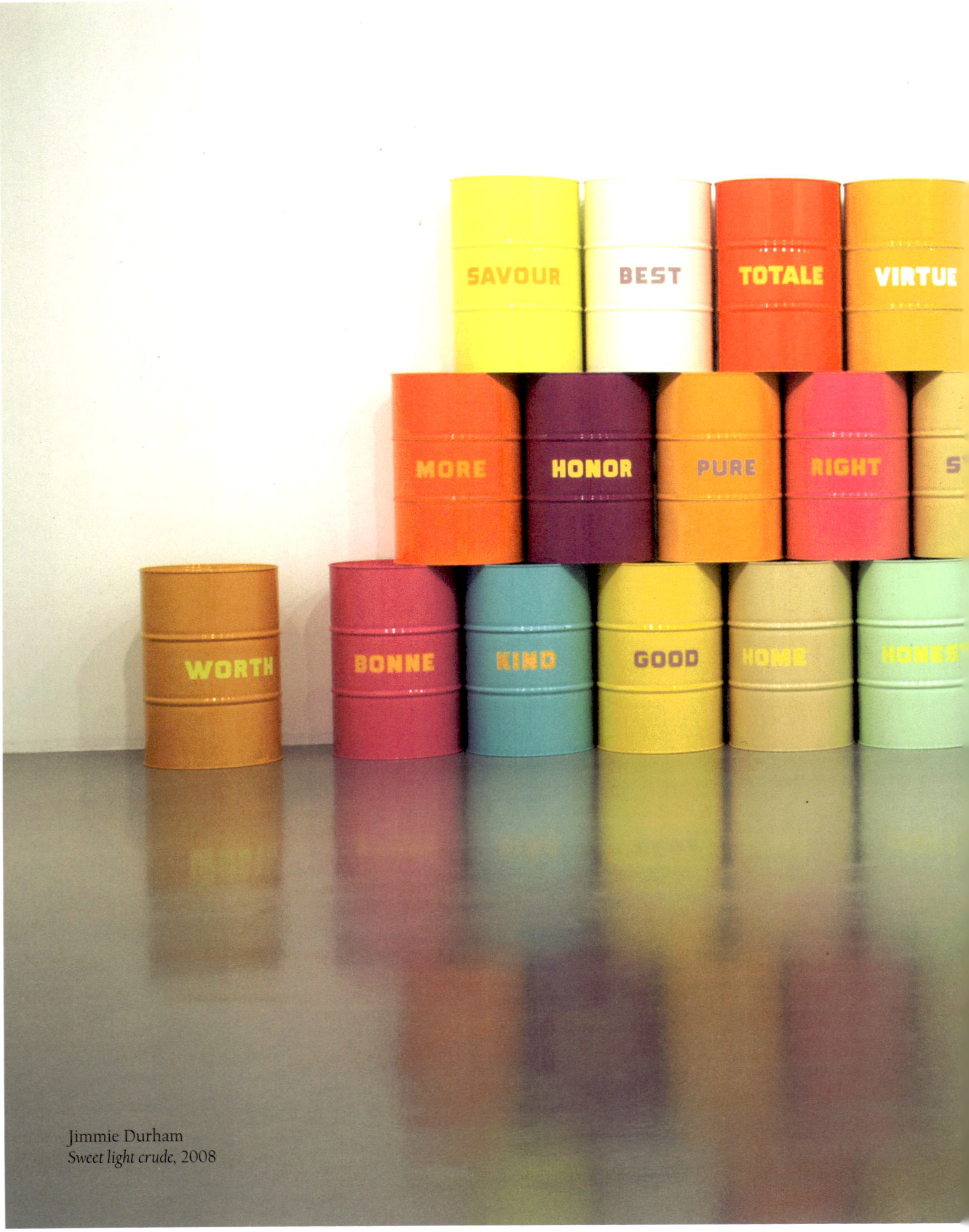

Jimmie Durham
Sweet light crude, 2008

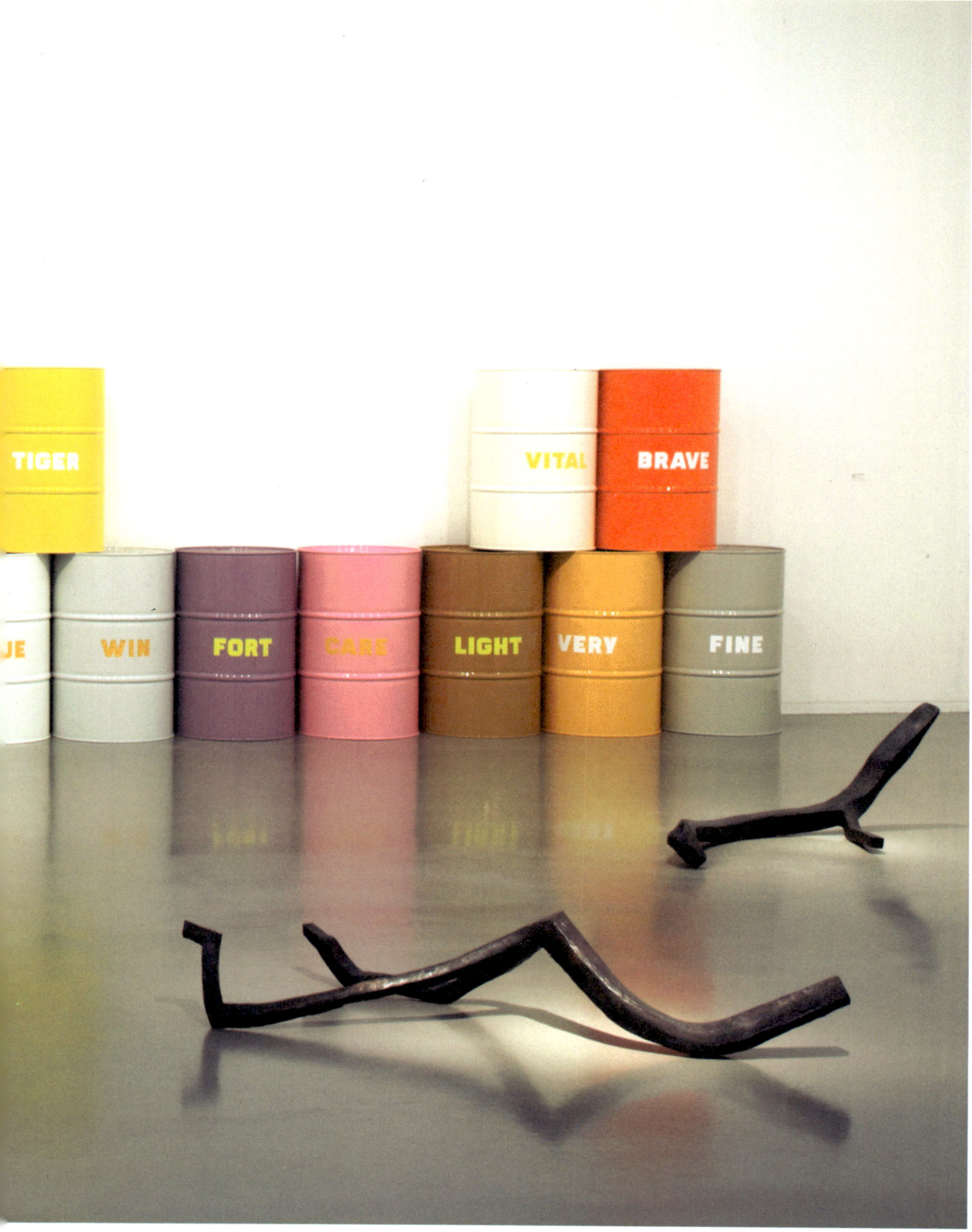
TIGER
VITAL
BRAVE
WIN
FORT
CARE
LIGHT
VERY
FINE

BENEATH A PETROLIFEROUS MOON

Jen Budney

we are the middle of 120
miles without services if we run
out of gas we are in super-deep shit*

NOT SO VERY LONG AGO, crude oil was so novel and plentiful that we were constantly looking for ways to integrate it into our everyday lives, but today most human beings in developed nations exist within an ecology of oil so complete that almost none of our needs and practices can be met without it. Oil has become a critical actor — if not *the* critical actor — in politics and culture worldwide. And where once we took our complicated dependence upon it for granted, today petroleum appears daily in the papers as an issue of contention and concern, from forecasts of peak oil to predictions of China's economic dominance, in the endless items on climate change and various, interminable wars, or in dreadful accounts of pipeline bombings and tanker blowouts. It is increasingly clear that we live and labour "beneath a petroliferous moon," as the poet Pablo Neruda so presciently put it in 1940.[1]

Petroleum has shaped and defined contemporary society more than any other substance. Without it, there would be no globalization, no plastic, limited transport, and a totally unrecognizable political landscape. Yet, in the developed world petroleum is for the most part extracted, refined, shipped, and consumed in near invisibility. Few of us truly understand the science, history, or prospects of the substance. As Brian C. Black, author of *Crude Reality: Petroleum in World History*, says:

> To varying degrees, all consumers of petroleum exist with almost no knowledge of the difficulty of locating, collecting, processing, and dispersing petroleum. Over more than a century and a half, the industry worked hard to insulate consumers from such details.[2]

There are many reasons for the petroleum industry's lack of transparency. The first global cartel, which eventually became known as the Seven Sisters, was formed in 1928, away from the public eye and government scrutiny, to control the production zones, transport costs, and sales prices of petroleum in an unstable free market.[3] Governments around the world have had to abide by the demands of the oil cartel to maintain their dominance or their basic security. Other nations and people simply suffer under the oil companies' activities and dictates. In their quest to maintain or increase profits, the petroleum industry has supported unpopular monarchies in oil producing nations, encouraged wars from which it could gain, and opposed the creation of OPEC (Organization of the Petroleum Exporting

*Unless otherwise noted, all lines of poetry in this text are taken from Marcella Durand's "The Anatomy of Oil," first published as a book by Belladonna Press in 2006 and reprinted in her book *AREA* (Brooklyn: Belladonna, 2008), 59–80. Printed here in full, pages 50–65.

Countries). It has sidestepped environmental regulations and human rights, encouraged the corruption of third world governments, and covered up when disasters happen away from Western media. And in the meantime, the petroleum industry has made consumers in the wealthy West very, very happy.

Most of us don't even understand the natural facts of oil. For example, did you know that a gallon of gasoline derives from plant material that decomposed and was pressurized under the earth's surface for thousands of years? And that same gallon is equivalent to ninety tons of plant matter, which is, to put it visually, like forty acres of wheat? One simple gallon. Imagine the gas as landscape the next time you fill up your car.

Here's another bit of trivia: according to industry sources, the oil and gas sector employs 9.2 million people in the United States alone.[4] So you could also reflect on your gas as capital, as the product of however-many man-hours of labour in the oil fields, in refineries, along transportation routes, and at your local gas station by engineers, chemists, earth scientists, and geologists, rig workers, refinery technicians, truck drivers, and service station attendants. Take a moment to conjure mental images of the faces of all those individuals whose efforts have led to you filling up your tank, and say "hello."

And while you're there: how did you choose your gas station today? Was it a simple matter of proximity, or do you have a favourite brand? Gasoline, like most commodities, is also heavily marketed, and most of us are vulnerable to marketing one way or another. Perhaps you possess a Shell Mastercard® and use it to collect CashBack. Or you are encouraged by Chevron asking us to "think differently about energy." Then again, you might be one of the millions of people drawn to the five-pointed red star Texaco logo, as evidenced by the quantity of retro-chic Texaco collectibles sold on eBay. *ABC News* reported that customers regularly pay more for brand name gas like Mobil or Shell, although gas from 'no-name' or generic service stations is indistinguishable from brand names.[5] What gasoline branding strategy engages you?

My point is: although we normally approach the petroleum industry in simple terms of economics (the price of a barrel of oil, or a gallon of gas) or ecology (air pollution, spills, and global warming), there are myriad ways to think about it. In the preface to his informative book *Oil: A Beginner's Guide*, Professor Vaclav Smil admits: "Squeezing the universe of oil into 60,000 words of text was an exercise in truncation and exclusion."[6] Marcella Durand's book-long poem "The Anatomy of Oil," which I quote from in this text and which is reprinted in this catalogue, is an extensive exploration of human histories and futures imagined through the mineral kingdom, upon which our whole existence is predicated. Although most aspects of the oil industry are as invisible to our eyes as vapours, the 'black stuff' underwrites our entire modern world, particularly in Canada. In his introduction to this publication, Smil explains that "In Canada, our relationship to oil is more important and more existential than in most other affluent nations, and not just because we produce and export a great deal of it...."[7] The distances we daily traverse within Canada in private vehicles and airplanes, our transport trucks, and our 'snow bird' travel to warmer climates all contribute to a per capita usage of oil that is more than twice the amount of the average

German's, and greater than one tonne more per year than a Swede's.[8]

In solubility how is water like oil
one is invisible and the other obscure
one combusts and the other is flammable
when one is put into the other it spreads
to where we stand watching how is one
like another how are you liking me how
are you where are you how are we

Beneath a Petroliferous Moon was an exhibition of contemporary art works on the subject of the petroleum industry. Surprisingly few artists have ventured to engage with this matter, although, as I have established, it presents an abundance of possibilities. In Canada, where a great deal of the writing about the oil industry is polemical,[9] the intention of the exhibition was not to persuade audiences to adopt a particular stance on oil, but rather to pique their curiosity and inspire them to learn more about the industry as it operates globally. Facts themselves are political, so there is no need for an ostentatious display of outrage; besides, as many of the artists suggest, while aspects of the industry may ignite protest or festivities, oil in its totality can only be approached with ambivalence, since without it our lives — with the bad and the good — are unimaginable. That said, it would be disingenuous not to acknowledge that most of the artworks in this show are more critical or apprehensive than celebratory. Where form becomes interpretable as commentary, the art works tend to express disapproval of a variety of industry practices.

We drift lower,
thirsty and surrounded by minerals,
the history of that, which:
that, which leads us to deep
histories, depths of hunger
for which there is no resolve

There is an ironic circuitry to an art museum exhibition about oil. After all, one could argue that the North American art museum is a direct product of the oil industry. Andrew Carnegie, the steel magnate whose fortunes rose on supplying the derricks and train tracks needed for the first oil boom in Pennsylvania, was the author of *Gospel of Wealth* (1889). He was a major funder of libraries and community centres across the United States and in Canada, and he was instrumental in shaping the discourse around education in art museums. John D. Rockefeller, Carnegie's colleague and the famously hard-nosed founder of Standard Oil Company, not only revolutionized the petroleum industry, but also defined the structure of modern philanthropy, establishing the Rockefeller Foundation in 1913, which is today a generous funder of the arts. 'Big oil' companies, particularly British

Petroleum and Shell, are frequent sponsors of art exhibitions; for example, BP has been an ongoing sponsor of London, UK's Tate Gallery for over two decades now, and a former BP chief executive, now CEO of a fracking company, is chair of the Tate Gallery's board.[10] It

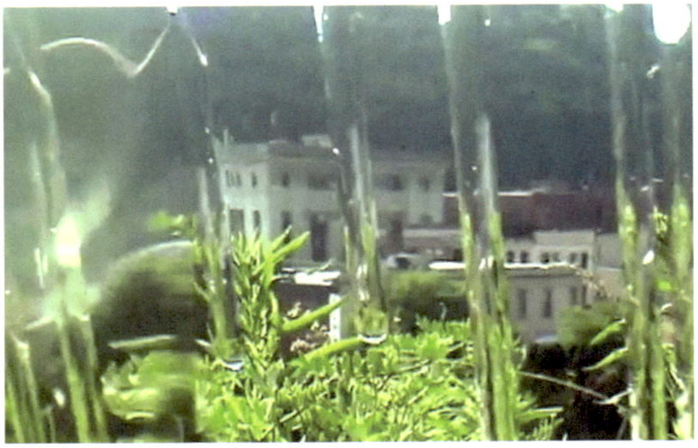

Robert Ladislas Derr,
still from the video *Oil City*, 2007

should follow as no surprise that there exist several activist groups dedicated to ending the sponsorship of art museums by oil companies.[11]

One work in *Beneath a Petroliferous Moon* touches peripherally on the interconnection between the arts and oil: Robert Ladislas Derr's video, *Oil City* (2007). This simple and elegant video loop focuses on the Oil City National Bank building in Oil City, Pennsylvania, a place that was once the heart of the world's first oil boom and host to the central headquarters of Pennzoil, Quaker State, and Wolf's Head Motor Oil Companies. Indeed, the richest of the American oil barons used to keep their money in the Oil City National Bank, but as the oil wells in the vicinity dried up, business moved elsewhere. Oil City fell into a depression, and the bank building itself was abandoned in 2001. Today, municipal leaders are attempting to remarket the city as a good place for artists to live and work — rent is cheap, after all. Of course, this is not what Rockefeller had in mind, but there is always a bust side to the boom of oil.

As it spreads along the sills impermeable
in best-intentioned effusion of life,
teeming, but between intent and discovery,
a canyon, chasm,
that is, to explore one seam
with intent to destroy the next

Beneath a Petroliferous Moon is not the first exhibition of visual art about oil — artists have been responding to the petroleum industry since Pablo Neruda penned his prescient poem in 1940, *Standard Oil Co.*, from which the title of this exhibition was taken. In 1975, for example, Hans Haacke referenced an oil company executive in his groundbreaking work, *On Social Grease*, about corporate sponsorship of the arts, and he continued to commentate on this subject in works such as *MetroMobilitan* (1985), which critiqued Mobil Oil's funding of New York's Metropolitan Museum of Art. In Canada in 1988, artist Rebecca Belmore staged the performance *Artifact #6718* to draw attention to Shell Oil's duplicity in sponsoring a Calgary, Alberta exhibition of Aboriginal art and artefacts while simultaneously despoiling the lands of the Lubicon Cree in the province's north. In recent years, particularly in the US, there have been a few group exhibitions focused on petroleum and automobile dependency, and we have seen interest in art about oil expressed in the publication of *Fuel* (2008), an issue of the hardcover art and politics periodical *Alphabet City Magazine*.[12] But *Beneath a Petroliferous Moon* was, to my knowledge, the first art exhibition about petroleum in Saskatchewan, Canada's second-largest oil producing province, a place that is at this very moment poised to begin extracting bitumen from tar sands,[13] as already practiced in neighbouring Alberta.

As a survey of recent artistic responses to the petroleum industry, *Beneath a Petroliferous Moon* intended to bring a glimmer of visibility to the multiple dimensions of the business of oil as it functions internationally. In developing the show, I hoped that the rich variety of global perspectives presented by the artists could assist in the public debate in our province about local tar sands development and human reliance on oil. The art works in the exhibition are, each in their own way, visually awe-inspiring and also devastating, testimony not just to the awesome and awe-ful nature of the subject but also to the dedication of the artists in facing their subject matter with honesty and ambition. The works' compelling aesthetics provided a gateway for thousands of visitors to the Mendel to learn more about the systems through which petroleum runs, and which are often run on petroleum: industrial, economic, ecological, and narrative.

Lines extend from continents
to seas unseen in this chasm.
Walls block escape and limit
sky, but here is where
we thirsty in the seams
liquid with end in sight,
are everywhere around.

No Canadian exhibition of art about petroleum would be complete without the inclusion of the works of Edward Burtynsky, the Toronto-based photographer whose sweeping views of landscapes altered by industry, including mine tailings, quarries, and scrap piles, have made him a household name in countries around the world. Burtynsky is an outspoken environmentalist, but has been criticized at times for making images of completely desecrated landscapes that are, in a word, sublime. Yet this is where their power lies, in their

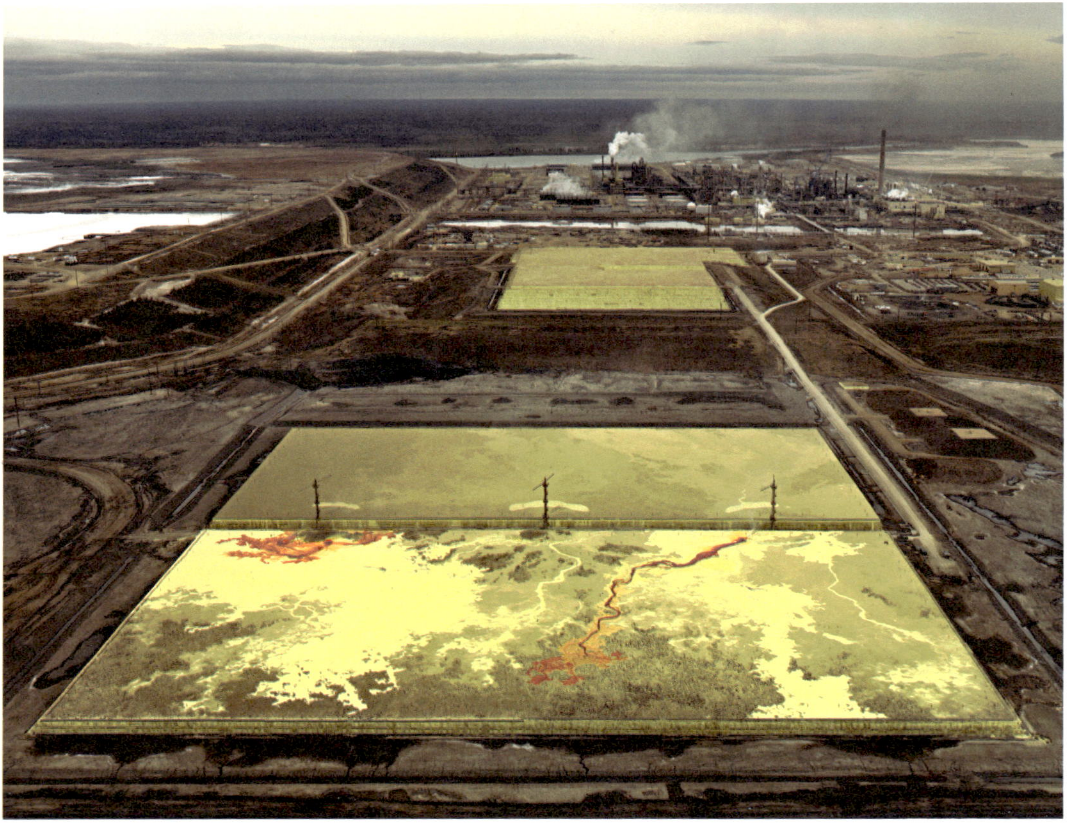

Edward Burtynsky
Alberta Oil Sands #6, Fort McMurray, Alberta, 2007

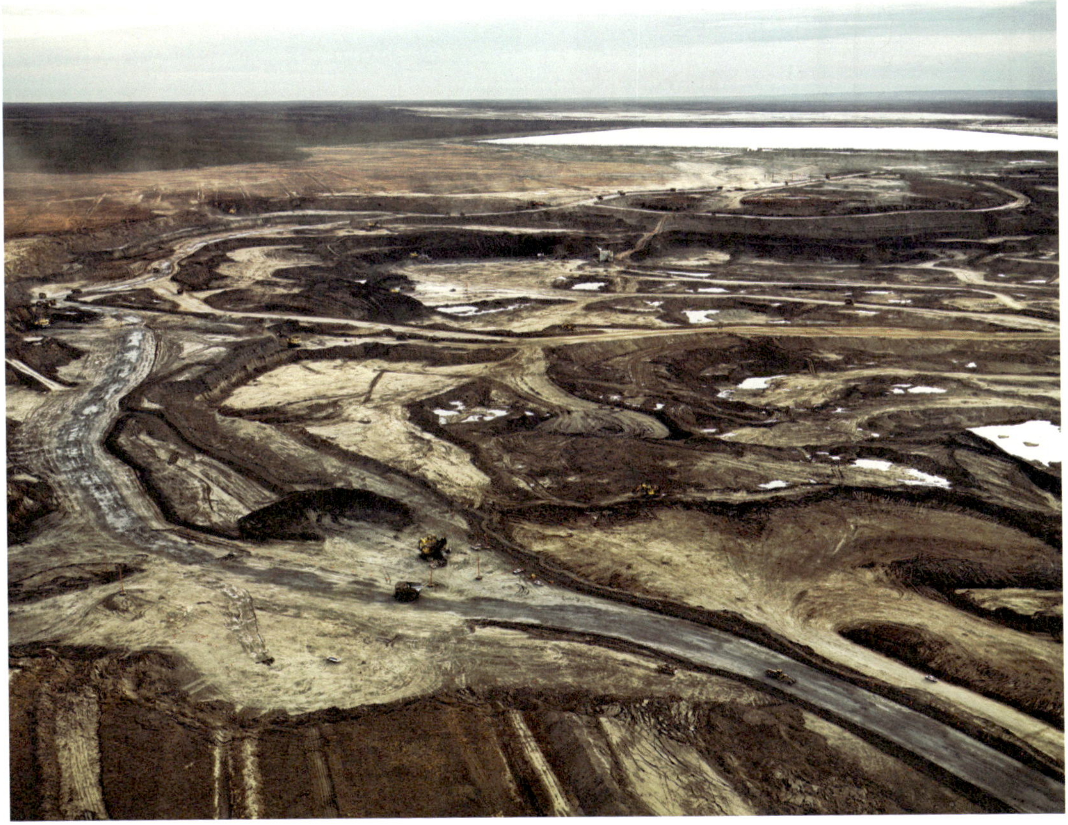

Edward Burtynsky
Alberta Oil Sands #5 , Fort McMurray, Alberta, 2007

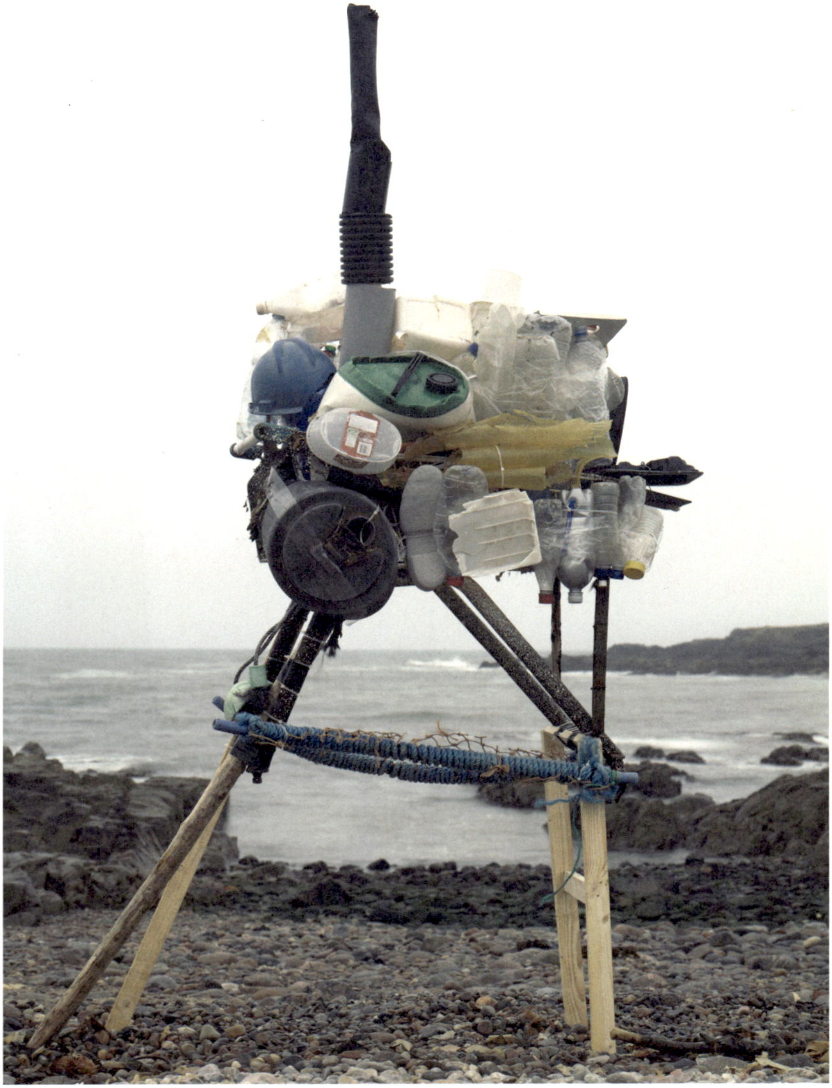

Ernst Logar
Woodside, 2008

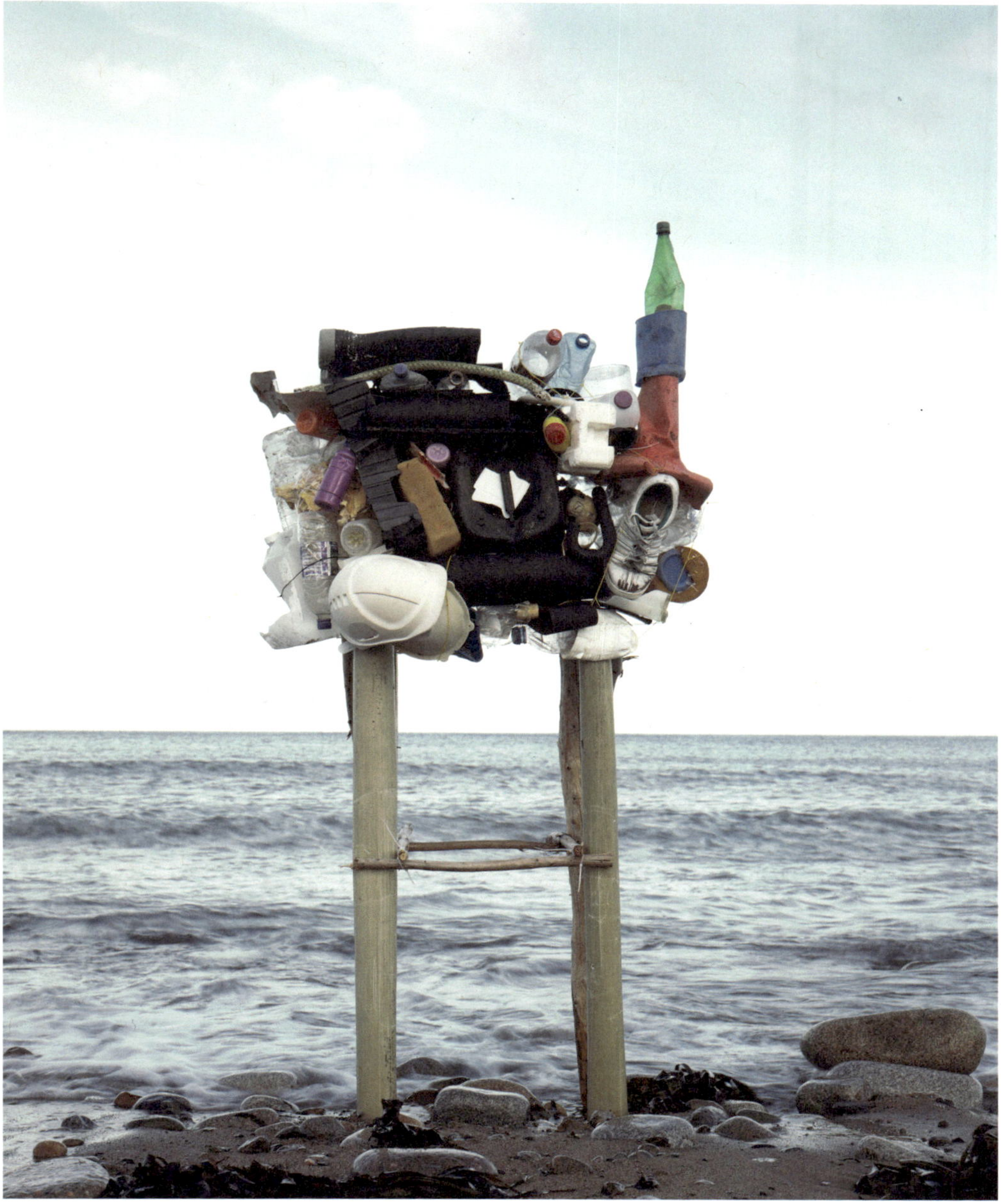

Ernst Logar
Torry, 2008

ability to elicit in the viewer awe, and then shock. His comprehensive *Oil* series, which he began in 1997, includes the subseries *Extraction & Refinement, Detroit: Motor City, Transportation & Motor Culture*, and *The End of Oil*. In the scores of images included in these series, we see mindboggling bird's eye views of tangled freeway systems, automobile plants resembling ghost ships, working and abandoned oil fields in California and Azerbaijan, respectively, with derricks (in the latter case, all decrepit) stretching to the horizon, and the utterly violated landscapes of northern Alberta, where tar sands excavation has scraped away the earth's surface and left pools of stinking bitumen everywhere. These are amongst the most important works of art made today about the environment, because they have moved so many people and influenced our perceptions of an industrial activity that few individuals outside the industry have been able to witness firsthand.

In her short essay on Burtynsky's work, Rebecca Solnit writes: "Facts themselves are political, since just to circulate the suppressed and obscured ones is a radical act."[14] In this vein, Austrian artist Ernst Logar spent three months investigating and documenting the oil industry in Aberdeen, Scotland. There are approximately 450 drilling platforms in the North Sea, making it the world's largest offshore oil region. In the late 1960s, the industry's establishment in Aberdeen made it the "oil capital of Europe," but since production peaked in 1999, the city's economy has begun to falter. Under these circumstances, Logar attempted to document the key operation centres of the Aberdeen oil industry and the effects of oil on the local economy. The resultant project, *Invisible Oil* (2008), includes hand-drawn maps of the oil and gas pipelines threading through Scotland, sculptures made from plastic detritus washed up from distant oil rigs onto Scottish shores (images of which were included in this exhibition), and copies of correspondence between the artist and representatives of many oil companies operating in the area, nearly all refusing him access to their offshore and on-shore properties for the purposes of visual documentation. Logar was in search of images of an offshore rig's wellhead bay, a laboratory where crude oil is analyzed, and an immersion room, where exploration is simulated. Instead, he received a stack of rejection letters, but he did manage to manipulate an official sign at the entry to an oil company site to read, "INVISIBLE OIL. Entry or work in this area is strictly FORBIDDEN," mimicking the industry's lack of transparency and, some would add, its subsequent lack of accountability.

The American journalist Peter Maass attempted to bring visibility to the murkiest regions of petroleum production, including Saudi Arabia, Equatorial Guinea, Venezuela and Iraq, in his informative and compelling book *Crude World: The Violent Twilight of Oil*. He began with the premise that "(o)ne of the ironies of oil-rich countries is that most are not rich, that their oil brings trouble rather than prosperity."[15] This is nowhere more true than in the Niger Delta, in eastern Nigeria. Maass travelled there to document the shoddy environmental standards of oil companies operating in this region, where emissions of sulfur dioxide and nitrogen oxide are so severe that "acid rain eats through sheet-metal roofs," and the locals suffer from a host of health problems, including "renal and cardiovascular failure, cancer, leukemia, emphysema, bronchitis, immune-system dysfunction and reproductive disorders."[16] It was under these circumstances that the non-violent Movement for the Survival of the Ogoni People struggled against the multinational Shell and the Nigerian military dictatorship, which led to (but did not culminate in) the November 1995 execution, by hanging, of the movement's leader, the writer Ken Saro-Wiwa.[17] The incident horrified people around the world and brought a new level of visibility to oil company collusion with dictatorial regimes, yet the region remains a very troubled area.

Bordering the Niger Delta to the east is the nation of Benin, home of the artist Romuald Hazoumè. Benin is not a producer of oil, but its proximity to the Niger Delta has led to a large trade in illegally siphoned oil, and this trade is the main subject of Hazoumè's artistic project. *Beneath a Petroliferous Moon* featured several of his 'masks' fashioned from discarded jerry cans and other objects found on the streets of Porto-Novo and Cotonou. Wryly humorous, the masks mock Western perceptions of African art, which were solidified during the period of early modernism in Europe, when Picasso, Braque, and others were so taken by the ceremonial objects brought back from colonial expeditions. More importantly, the masks comment with droll insight on the continuing process of capital exchange between the western world and modern Africa. Africa is inundated with garbage from the West; the West moves in, extracts the wealth, and leaves behind its detritus. But Hazoumè resists this burden, this economic and environmental oppression, by transforming the West's garbage into items of real value to be consumed primarily by people living in the West. His objects also tell the story of the daily struggle for survival faced by so many of his compatriots, "of market women who balance the [jerry] cans on their heads, or people on mopeds who pile them up to make bizarre structures and sometimes fly through the air with them after accidents."[18] The artist estimates that about 90 per cent of the Beninese buy smuggled rather than legal oil, as legal petroleum products are too expensive. "If the supply of gasoline coming from Nigeria in these containers were to be stopped, we'd all be finished!" he explained. "The country and the government would collapse. The energy coming from the cans is the force of the people."[19] The masks honour the people of Benin, whose power and resistance is derived from the energy they carry in jerry cans.

Like Hazoumè, Canadian artist Brian Jungen first received international acclaim for 'masks' fashioned from consumer-culture items — in his case, Nike Air Jordan trainers, which were transformed into semblances of traditional masks from the Aboriginal Northwest Coast (*Prototypes for New Understanding*, 1998–2005). His work in *Beneath a Petroliferous Moon* is made from a jerry can, but not a battered, used one like those employed by Hazoumè. *Dragonfly* (2008) is made from a new, orange/red plastic jug, the kind one finds everywhere around Fort St. John, British Columbia, Jungen's hometown, the official centre of B.C.'s oil and gas industry and seat of the province's Oil and Gas Commission, and closest city to the Doig River Indian Reserve, where his mother's people live. In Fort St. John, jerry cans are used to fill up snowmobiles and chainsaws, and they carry emergency supplies of gas for personal vehicles, since the region is, ironically, short on gas stations. *Dragonfly* is one of a series of jerry cans on which Jungen has made drawings of northeastern B.C.'s poisonous plants and endangered insects by drilling hundreds of tiny holes. Like Hazoumè's masks, these sculptures comment on the impact of the oil industry on indigenous economies and

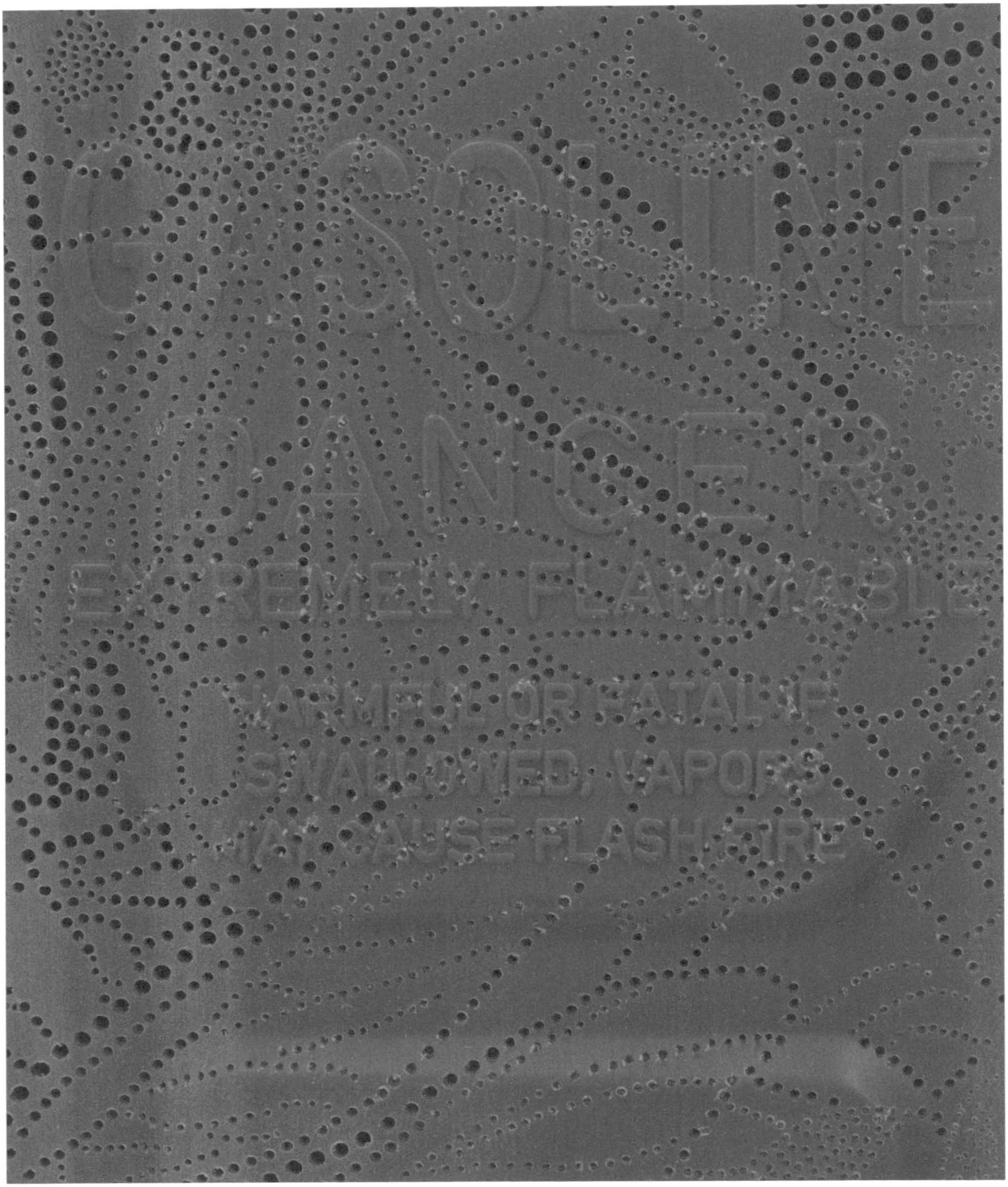

Brian Jungen
Dragonfly (detail), 2008

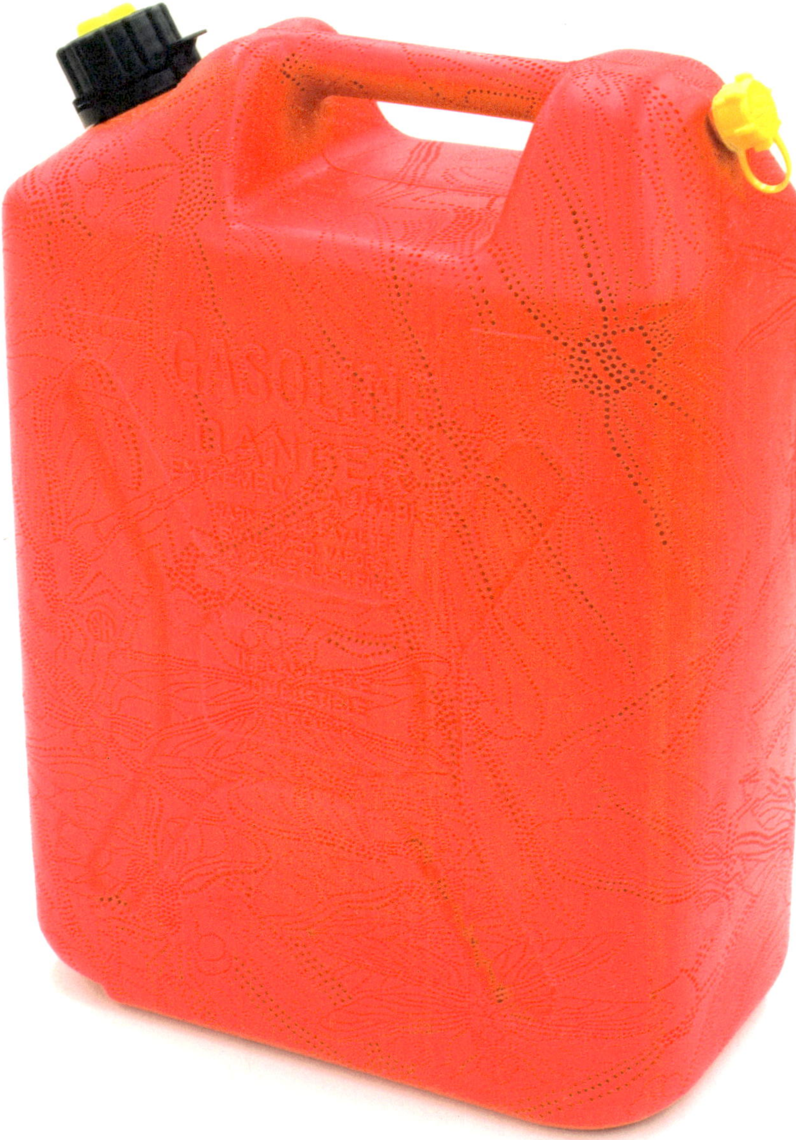

Brian Jungen
Dragonfly, 2008

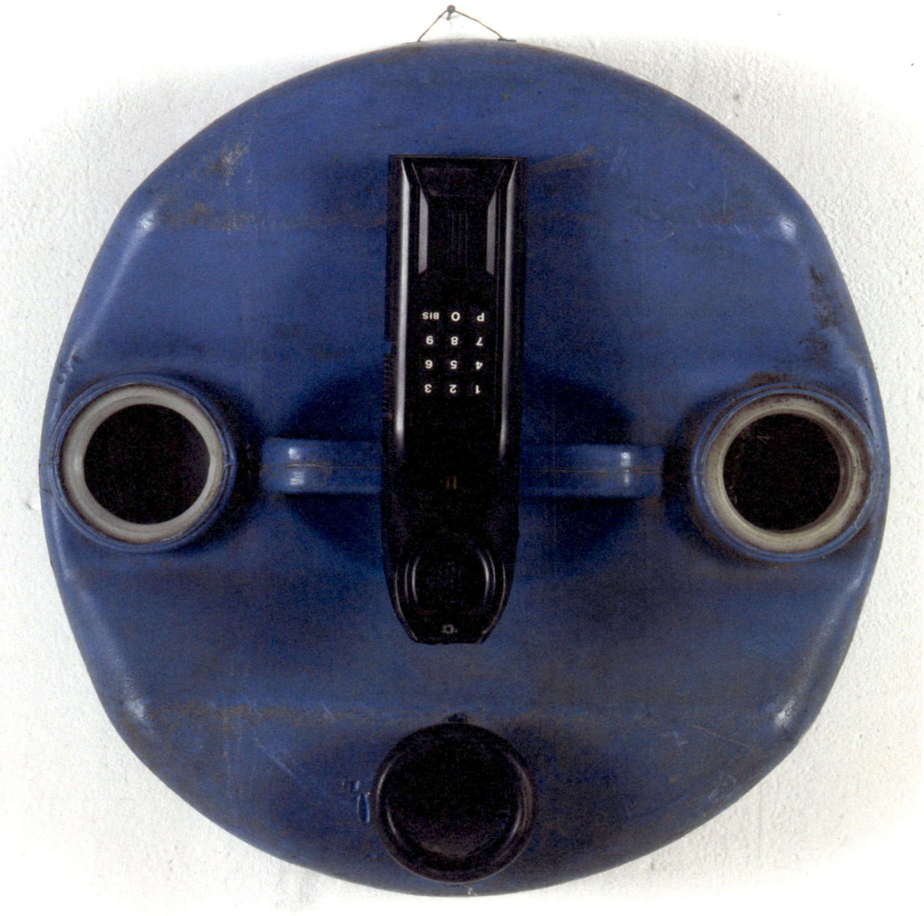

Romuald Hazoumè
Dan, 1992

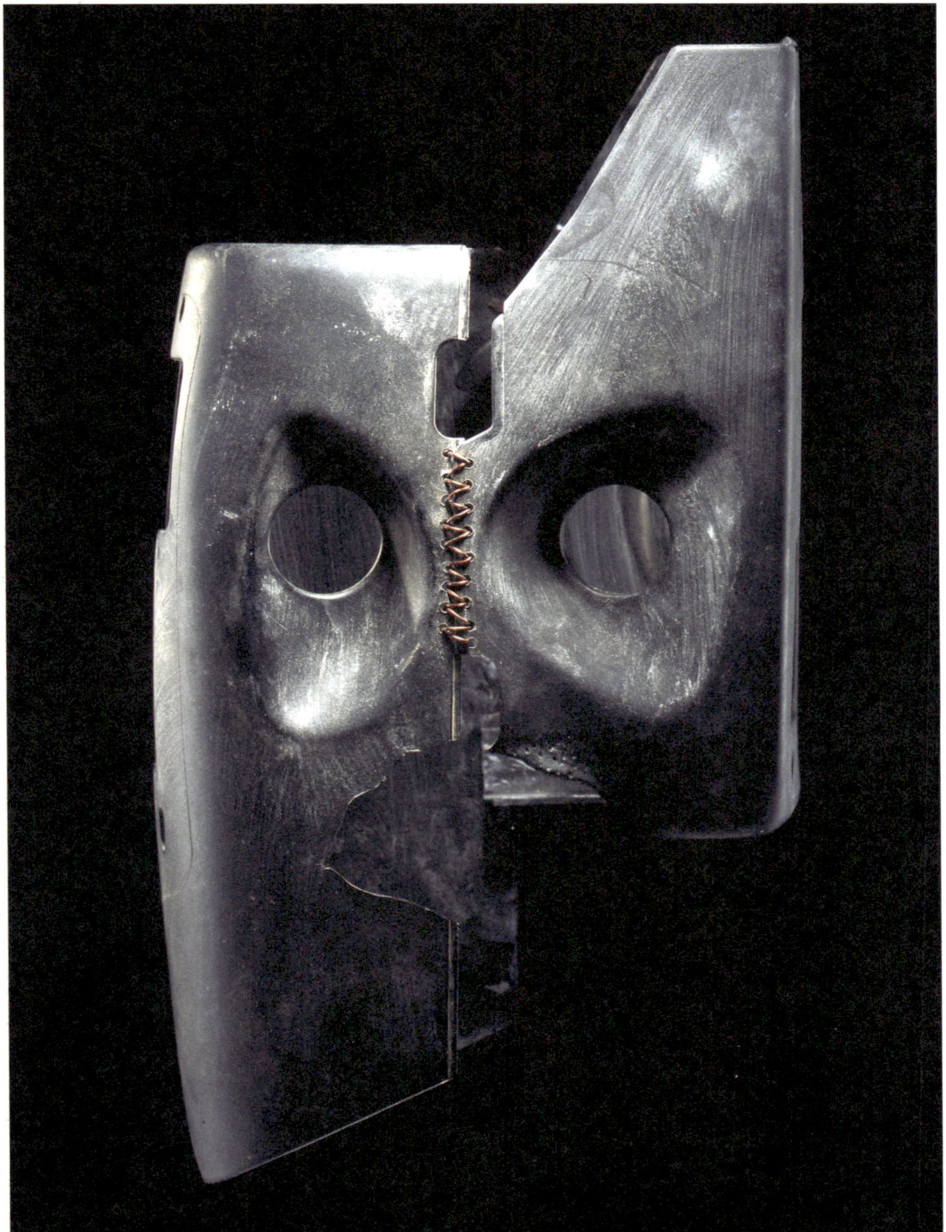

Romuald Hazoumè
Black 2020, 2005

culture, while simultaneously preserving and presenting aspects of tradition. In this case, Jungen's drilled drawings resemble the traditional Dene bead designs his family sewed onto animal hides. The metamorphosis he enacts upon the jug recalls life on reserves, where discarded objects are often converted, recycled, or 'upcycled' into other usable forms as a solution to the local lack of commercial or financial resources.

> if gravel could be ground and rock extracted
> sand crushed and siphoned, flyaway soil captured
> snow and ice melted, rocky beaches compacted, and
> beneath the giant thumping machines small drops well
> and conglomerate into solid puddles

Where Burtynsky, Logar, Hazoumè, and Jungen have made works that deal, in different ways, with the known effects and activities of the oil industry, American Louisa Conrad has produced a stunning collection of research on the uncertain future of the Beaufort Sea, the marginal stretch of sea in the Arctic Ocean where the mighty Mackenzie River empties, north of the Yukon and the Northwest Territories. The Beaufort Sea contains major gas and petroleum reserves beneath the seabed, and is already home to about 200 offshore wells that have displaced barite clay, caustic soda, and all sorts of heavy metals into the water. Conrad visited the area in 2009 to explore the controversial Mackenzie Gas Project, a proposed 1,196-kilometre natural gas pipeline system along the Mackenzie Valley, which would connect the offshore gas fields with southern markets and result in a massive increase in drilling activity and, therefore, a local economic boom. During her six weeks in the region, Conrad documented the multiplicity of voices and perspectives on the Mackenzie Gas Project proposal. As Logar did in Scotland, she collected industry maps and information on existing pipeline infrastructure, as well as geological surveys of permafrost and species migration routes. She made drawings of threatened local flora and attempted to journey by boat nearly 200 kilometres through the Mackenzie Delta to the mouth of the Arctic Ocean, where she would have been in view of the Kendall Island Bird Sanctuary. One of the videos included in her presentation in the Mendel exhibition documents her view of the sea and her conversation with her Inuvialuit guide when their boat became stuck in miles of silt and they were forced to push it north together for more than eight hours. The moody, uncertain sky, the shifting directions of the wind, their anxious voices, and their seemingly fruitless toil became metaphors for the Mackenzie Gas Project controversy as a whole, and what Conrad described as its "essential irresolvability."[20] There is a quiet stillness to all of Conrad's work, a stillness that beckons viewers to look carefully and play close attention. I read this as a metaphor for our need to reflect deeply on the future of the Beaufort Sea, even though it is a place that most of us will never know first-hand. The artist still maintains a blog on this project ("Unfinished Bridge"), while the prospects of the Mackenzie Gas Project and the Beaufort Sea environment remain uncertain.

Conrad also used video to document the Syncrude Tar Sands Processing Plant and Tailing Ponds in Fort McMurray, Alberta, as it would be there that Beaufort Sea gas would be piped. Fort McMurray has been called the Hydrocarbon Capital of Canada. It is accessible by plane or by driving north from Edmonton on 'Hell's Highway' or the 'Highway of Death,' a section of Highway 63 known for its congestion, perilous speeds, and intoxicated drivers. Fort McMurray itself is known for its social problems, which range from overstretched infrastructure to soaring drug use. It is a place UK-based artist Susan Turcot has visited for her ongoing series of projects dealing with the economies of natural resources that are thought to be approaching crisis and that influence political processes globally. Her 2010 *Exxon-Valdez's Offspring* drawing series, exhibited at the Mendel, references all the oil spills that have happened since the infamous 1989 oil tanker accident off the coast of Alaska, spills that seem to be inevitable occurrences in the process of oil extraction and transportation. These deeply moody charcoal drawings are contemporary riffs on the landscape genre, evoking the endless extensions, crossings, and support constructions of transcontinental oil pipelines as a vulnerable arterial system. The drawings have a cinematic quality, similar to a Sergei Eisenstein montage, in which the sensational event of an oil spill has an even greater impact through the juxtaposition of dramatic close-ups.

Motifs from Turcot's drawings are also taken up in a two-armed mobile, which she assembled using found objects, including the core of a car tire and a palm frond painted black. This sculpture asks — in a hidden piece of text — whether Canadians are prepared to live in a petrostate. Its question, "Slavery?", evokes Pablo Neruda's diatribe against the workings of Standard Oil:

> However entangled the petroleum's arteries may be,
> however the layers may change their silent site
> and move their sovereignty amid the earth's bowels,
> when the fountain gushes its paraffin foliage,
> Standard Oil arrived beforehand
> with its checks and its guns,
> with its governments and its prisoners.[21]

Although politically urgent, Turcot's works are highly contemplative; they are not factual representations or documentations, but reflect her own reactions and thoughts about what she has seen. The artist's choice to work with the fossil resource of charcoal as a medium for her drawings is a deliberate reflection upon her own implication in the use and destruction of natural resources.

when sudden flooding
drives us lower past
crushed abyss of uncertain epoch; from the rims
falls darkly staining
excess

Louisa Conrad
The Last Arctic, I: Kendall Island Bird Sanctuary
& Site of the Proposed Niglintgak Barge-Based Gas Conditioning Facility
(Beginning of Pipeline), 2009

Louisa Conrad
The Last Arctic, II: Kendall Island Bird Sanctuary &
Site of the Proposed Niglintgak Barge-Based Gas Conditioning Facility
(Beginning of Pipeline), 2009

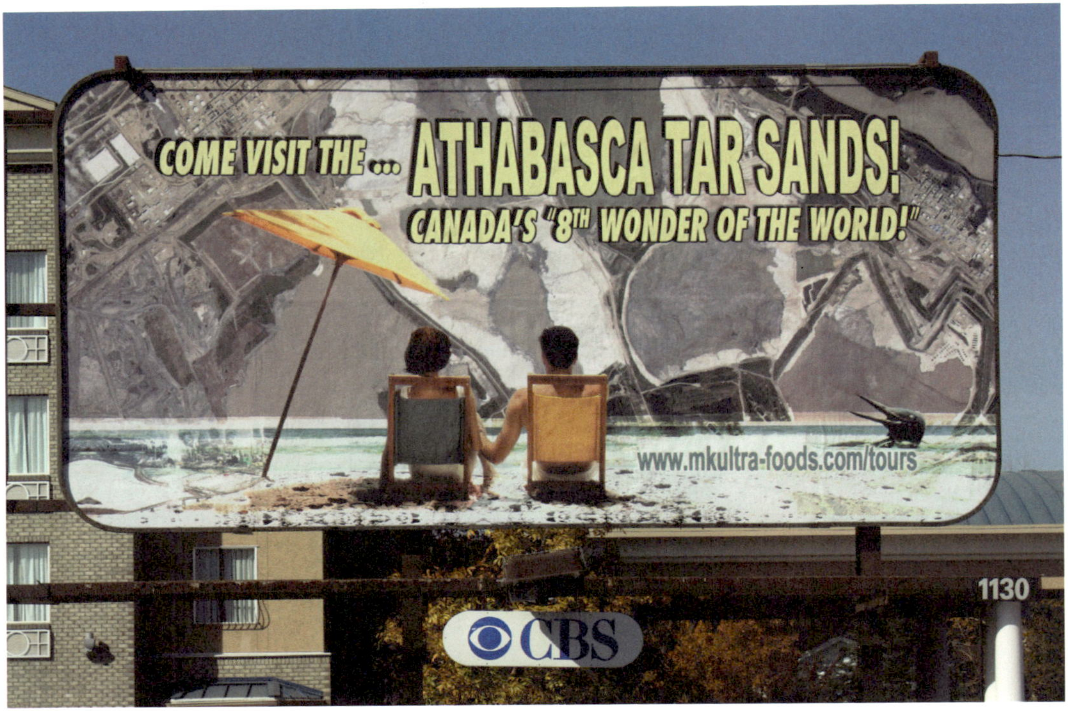

David LaRiviere
Come Visit the Athabasca Tar Sands, 2012

Robyn Moody's *Power 2: Heart Lake as seen through the eyes of Manley Natland* (2009) is a darkly glittering kinetic sculpture that pays homage to one of the most diabolical ideas ever devised for oil extraction. Manley Natland (1906–1991) was an American petroleum geologist who, in the 1950s, came up with the idea of extracting the bitumen from Alberta's tar sands by detonating underground nuclear explosions. His project was proposed to the Alberta government, which spent some time seriously considering it. The premise was that the heat generated from the explosions would easily separate oil from the sand underground, and it could then be pumped to the earth's surface. Fortunately, this project did not take hold (today, oil and sand are separated using 250 degree steam pumped into the earth through deep boreholes). Moody's sculpture takes the form of Heart Lake, which is a real lake that exists within Alberta's tar sands region, along with Hope Lake and other lakes represented in this sculptural series. The lake currently remains untouched by tar sands development, and Moody has chosen it because its current status and its name imply resistance. Made of hundreds and hundreds of tiny, rotating mirrors that catch and reflect the light from the gallery's ceiling, Moody's sculpture glistens and glitters blackly, as though it were a lake made of the darkest crude. For Moody, this sculpture is about "the madness and denial of reason that comes about through a single-minded quest."[22]

Saskatoon artist David LaRiviere also engages with a madness of ideas related to the tar sands. For *Beneath a Petroliferous Moon*, he created a public intervention in the form of a billboard located next to a major thoroughfare in Saskatoon's downtown.[23] *Come Visit the Athabasca Tar Sands* (2012) is from the artist's ongoing *Modest Proposal* series, which takes its name from Jonathan Swift's 1729 satirical essay that mocked elite Irish attitudes towards class by proposing that poor families alleviate their situations by selling their babies as food for the rich. In this piece, LaRiviere unpacks and ultimately mocks the 2006 statement by Canadian Prime Minister Stephen Harper that the tar sands project be considered "an enterprise of epic proportions, akin to the building of the pyramids or China's Great Wall. Only bigger."[24] That the excavations are of epic proportions the artist does not question — his billboard displays a bird's eye view of the devastated Athabasca Valley. Rather, LaRiviere questions the Prime Minister's comparisons of the tar sands to human projects that are considered wonders of the world for their beauty and durability. An image of beach-goers lounging under an umbrella, taken from a 1950s periodical, is superimposed on top of the view of the tar sands, creating an impossible, otherworldly image. Where a beautiful sunset might be are, instead, the tar sands. The billboard is plastered with an invitation to visit the site. Although the picture is unbelievable, the billboard is convincing, particularly given recent attempts by the city of Edmonton to lure Saskatonians to Alberta for weekend getaways, a campaign that has employed billboards of similar proportions. On the bottom of the billboard is a URL for LaRiviere's website (www.mkultra-foods.com), which contains a host of other interventions into and commentary on media. Visitors to the Mendel were directed to the billboard by a postcard bearing the billboard image on one side and a map to the destination on the other.

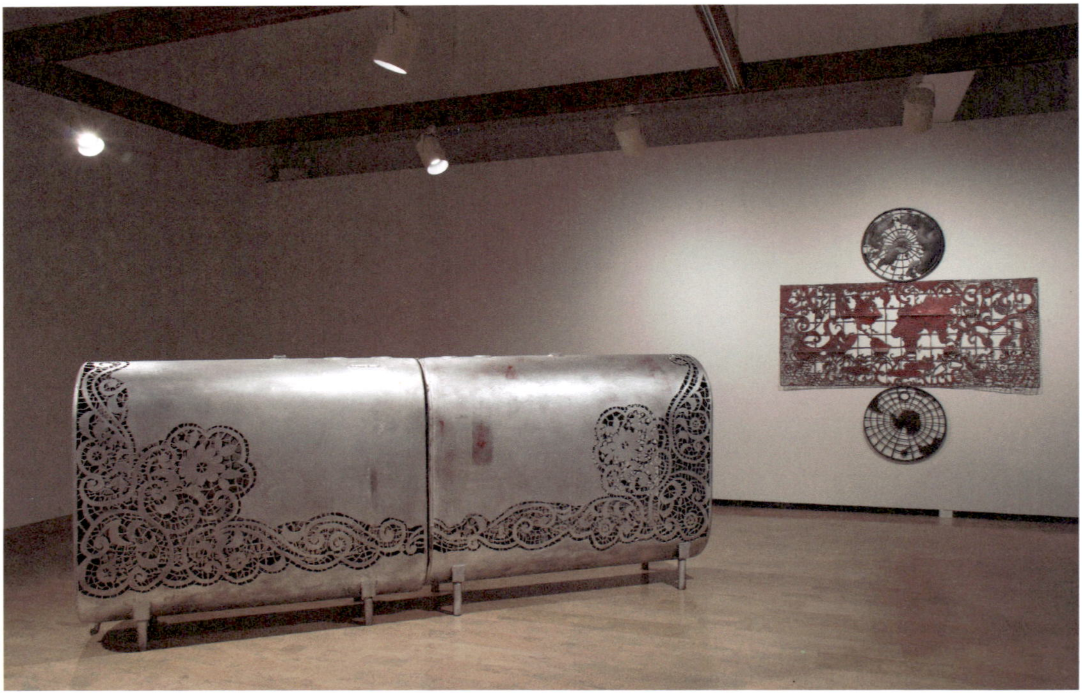

Cal Lane
Oil Fired Tanks, 2009
and *Oil Drum Skeletal Red Map*, 2011

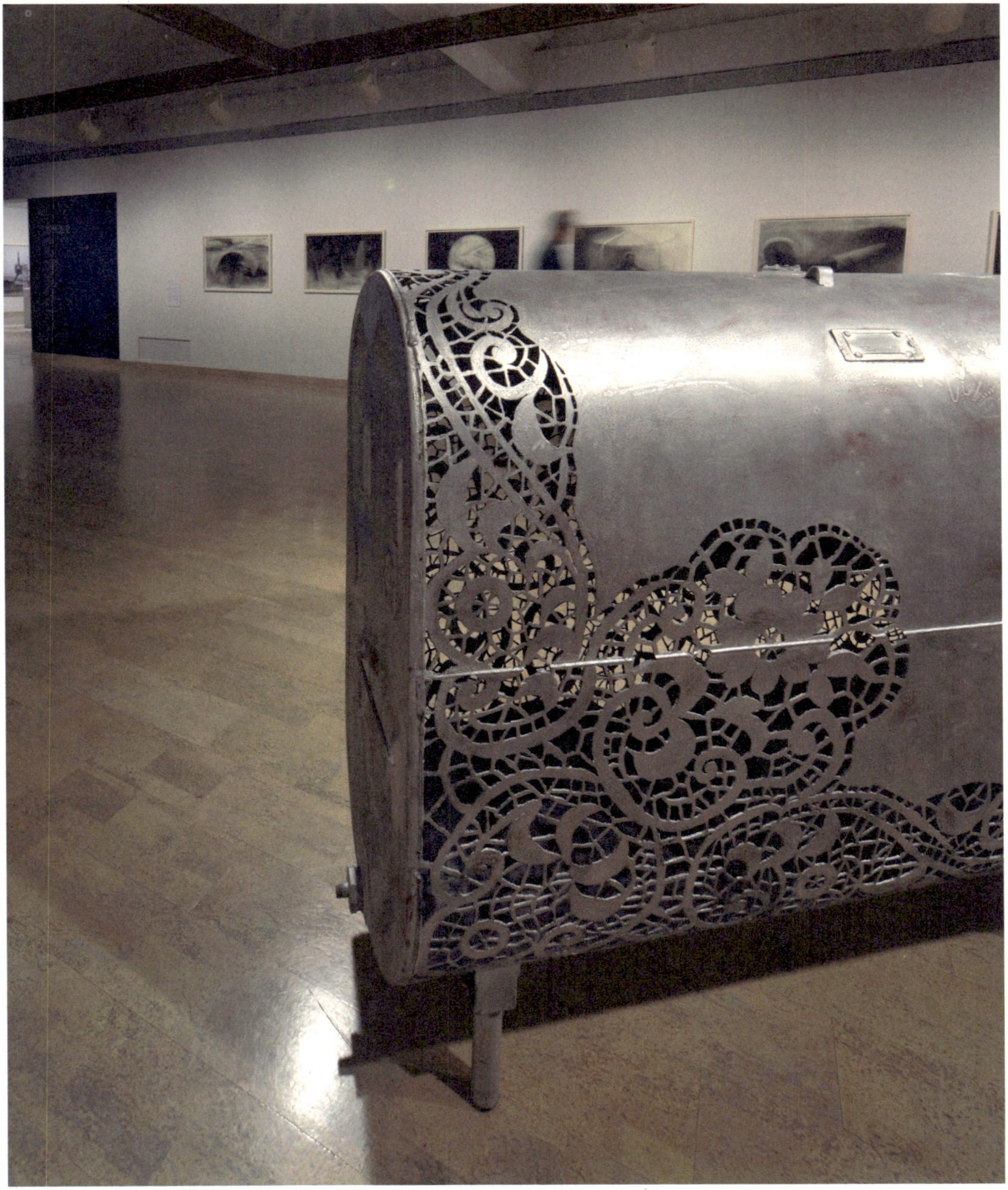

Cal Lane
Oil Fired Tanks (detail), 2009

If a comparison of the tar sands to the Great Wall of China seems rather euphemistic, the sculpture by Jimmie Durham suggests that such sugarcoating of hard realities is a longstanding petroleum industry activity. *Sweet Light Crude* (2009) is a sculpture of variable dimensions made up of real oil barrels painted in pastel colours. For his retrospective at the Musée d'art moderne de la Ville de Paris, Durham stacked 25 of these painted oil barrels, emblazoned with words like "worth," "kind," "home," "honor," and "virtue," along with "tiger," "vert," and "more." Durham has long used oil barrels in his work, partly because he likes their industrial form and partly because they reference the second expulsion of his people, the Cherokee, from the oil-rich lands of Oklahoma, where they had been settled only 50 years after being forced out of the Carolinas. Of course, real new oil drums often have words on them, such as "Shell" or "Total." *Sweet Light Crude* reflects upon the petroleum-industry classification for a particular grade of oil and upon the industry's penchant for a particular form of speciousness. In *Beneath a Petroliferous Moon*, a smaller version of the installation appeared, *Sweet Light Crude (Dubai-Oman)* (2008), with five candy-coloured barrels bearing the words "good," "true," "total," "pure," and "tiger."

its beauty, inverse to its fertility
but why call it beautiful? I can't even see it

it's a national monument, enclosed within
itself and meant to be seen in a sort
of grouping, a clustering together, a kind
of huddle.

Cal Lane also makes her work from oil barrels and oil tanks, but in her case the barrels and tanks are used and worn. She dismembers them, flattens or reshapes them, and incises delicate drawings and patterns into them using a blowtorch. Her final sculptures embody contradictions of a different sort than Durham's, juxtaposing industry and domestic craft, masculine and feminine, opacity and transparency. Yet, the title of her oil drum and oil barrel series is *Sweet Crude*, as the term applied to oil products embodies the great problematic that is our dependence on oil. But sweet crude is also the most sought-after form of petroleum; it is the kind of crude oil that nations have waged wars to acquire or to protect. The sculptures in Lane's *Sweet Crude* series indicate a relationship between God and oil, incised as they are with medieval Christian icons along with mythological beings, urban street scenes, and animal-studded landscapes. Her splayed-open barrel, transformed to a two-dimensional wall work, takes the form of a cross or Gothic cathedral floor plan. The clear reference to Europe's so-called Dark Ages poses at least two questions: What will become of our civilization when this oil runs out? And how will our fossil-fuel mad age be recalled by generations to come? In other words, are the next dark ages to come, or are we in them now?

Robert Ladislas Derr's video, centred on the Oil City Bank, is a continuous loop. Each loop endures for the length of time it takes a quart of motor oil to run over the lens of the camera. A quart of oil is a discrete quantity that functions in today's economic system akin to a banknote or to an ounce of gold. But like a dollar or an ounce of gold, oil is so much more. In its fueling of the modern world, oil shapes our very identities; it gives form and sustenance to our privileged ideas of what it means to be human, and it simultaneously underwrites many of our worst nightmares about our present and our future. Artists and poets have the ability to express the multidimensionality of this essential substance. What we do with what they show remains an open question.

if only we could stop the car if only we could look
*
at it closer if shoulder to
shoulder we stand closer together
but one would not be he
who has had nothing but plenty

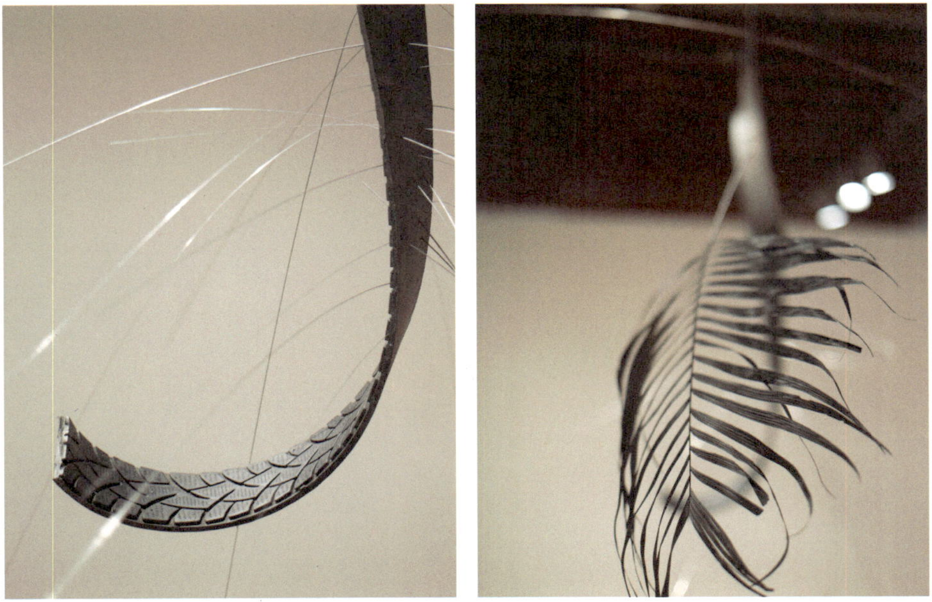

Susan Turcot, *Slavery I*, 2010, (details)

NOTES

1. Pablo Neruda, "Standard Oil Co." in *Canto General*, trans. Jack Schmitt, © 2011 by Fundacion Pablo Neruda and the Regents of the University of California. Published by the University of California Press, 176–177.

2. Brian C. Black, *Crude Reality: Petroleum in World History* (Maryland: Rowman & Littlefield, 2012), 5.

3. The group comprised Anglo-Persian Oil Company (now BP), Royal Dutch Shell, Standard Oil of New Jersey (Esso; now Exxon Mobil), Standard Oil Company of New York (Socony; now Exxon Mobil), Standard Oil of California (SoCal; now Chevron), Gulf Oil (now Chevron), and Texaco (now Chevron). For an excellent documentary on this subject, see Al Jazeera's four-part documentary, *The Secret of the Seven Sisters*, 2013. At the time of this book's publication, the documentary was available for free viewing on YouTube, http://youtube.com, and on the Top Documentary Films website, http://topdocumentaryfilms.com.

4. See www.api.org (American Petroleum Institute) for more statistics on oil industry employment. Keep in mind that some critics, such as those at the 'green' media outlet treeghugger.com, claim that these figures are overstated by as much as four times.

5. Elisabeth Leamy, "Generic vs. Brand-Name Gas: Are They Different?" *ABC News*, March 24, 2007.

6. Vaclav Smil, *Oil: A Beginner's Guide* (Oxford: One World Oxford, 2008), 10.

7. Vaclav Smil, this publication, 16

8. Smil, this publication, 22–23

9. See, for example, Andrew Nikiforuk's *Tar Sands: Dirty Oil and the Future of a Continent* (Vancouver: Greystone Books and David Suzuki Foundation, 2010) as well as Ezra Levant's *Ethical Oil: The Case for Canada's Oil Sands* (Toronto: McClelland & Stewart, 2010).

10. See Susanna Rustin, "Can Liberate Tate free the arts from BP?" in *The Guardian*, 24 April 2013. www.guardian.co.uk/artanddesign/2013/apr/24/liberate-tate-arts-bp.

11. See *Not If But When: Culture Beyond Oil* (London, UK: Art Not Oil, Liberate Tate, and Platform, 2011).

12. John Knectel, ed., *Fuel. Alphabet City* no. 13. (Cambridge, MA and London, England: MIT and Alphabet City, 2009).

13. See Claudia Cattaneo, "Hope lives for Saskatchewan's oil sands," in the *Financial Post*, February 12, 2013 (business.financialpost.com/2012/02/13/hope-lives-for-saskatchewan-oil-sands/). For the impact of tar sands development on Saskatchewan First Nations and Métis communities, see the film by Warren Carriou and Neil McArther, *Land of Oil and Water* (2009), which was screened at the Mendel Art Gallery on September 29, 2012, on occasion of the exhibition *Beneath a Petroliferous Moon*.

14. Rebecca Solnit, "Poison Pictures," in *Storming the Gates of Paradise: Landscapes for Politics* (Berkeley and Los Angeles: University of California, 2007), 136.

15. Peter Maass, *Crude World: The Violent Twilight of Oil* (New York: Vintage Books, 2009), 3.

16. Maass, 68.

17. See Ken Saro-Wiwa, *A Month and a Day in Letters* (Oxfordshire, UK: Ayebia Clarke, 1995).

18. Bartholomaus Grill, "Beyond Europe: Encounters with Romuald Hazoumè in His Native Country of Benin," in Romuald Hazoumè: *My Paradise — Made in Porto-Novo*, ed. Martin Henatsch (Neumunster, Germany: Herbert-Gerisch-Stiftung, 2010) 27. Individuals on mopeds can carry up to 500 litres of gasoline in large bunches of jerry cans tied together like giant balloons, transforming themselves into potential bombs. When accidents happen — and they do frequently — people go up in smoke simply because they want to feed their families.

19. Grill, 96.

20. Louisa Conrad, artist's statement, accessed June 18, 2013, www.louisaconrad.com/arctic/project-description/.

21. Pablo Neruda, "Standard Oil Co.", from Neruda's *Canto General*, translated by Jack Schmitt, © 2011 by Fundacion Pablo Neruda and the Regents of the University of California. Published by the University of California Press, 177.

22. Robyn Moody, artist's statement (2009), accessed June 20, 2013, vimeo.com/17110020

23. The billboard was located next to the railway tracks crossing Idylwylde Drive at 25th Street, facing south.

24. Paul Wells and Tamsin McMahon, "Oil power," *Maclean's*, April 2, 2012, 15.

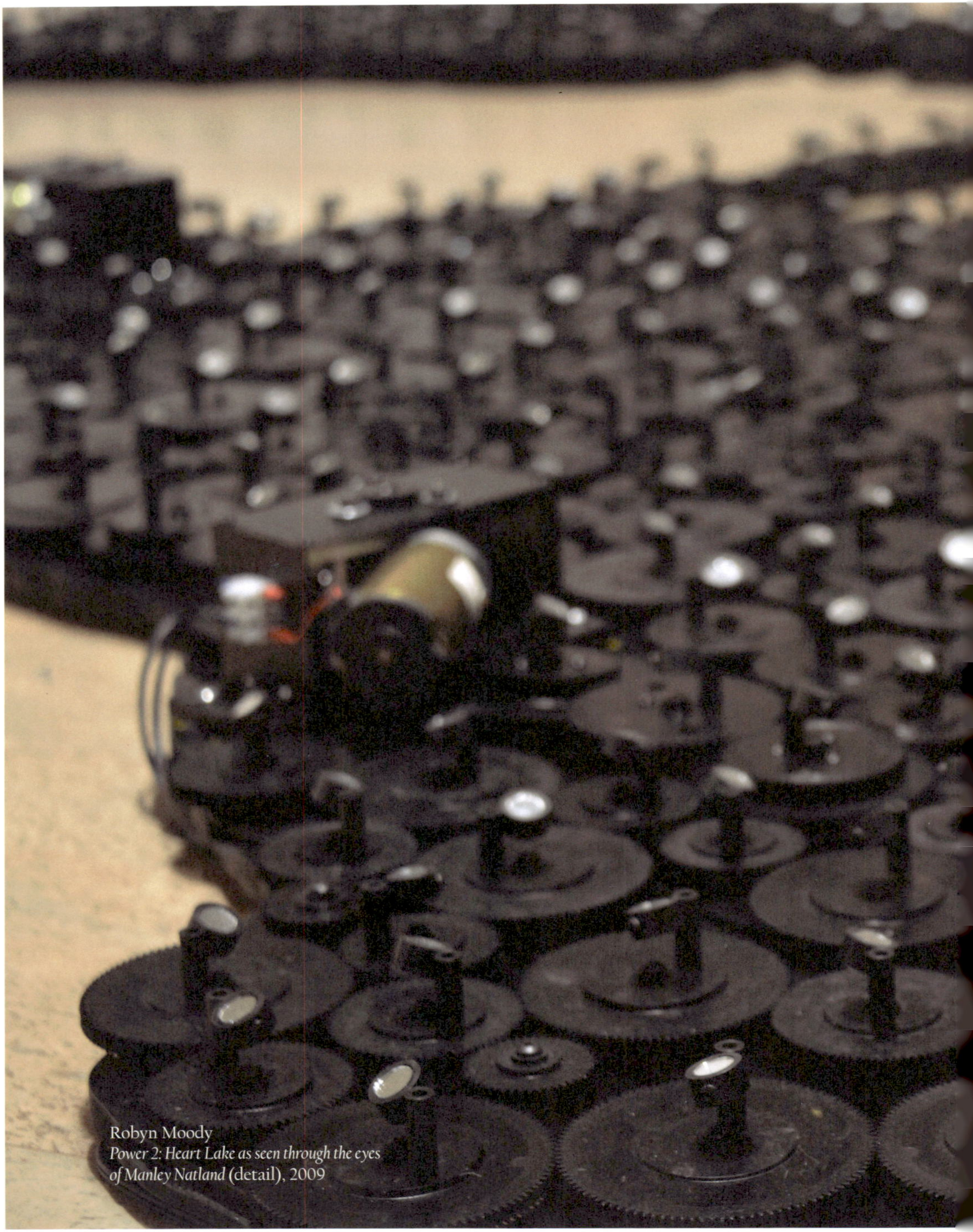

Robyn Moody
Power 2: Heart Lake as seen through the eyes of Manley Natland (detail), 2009

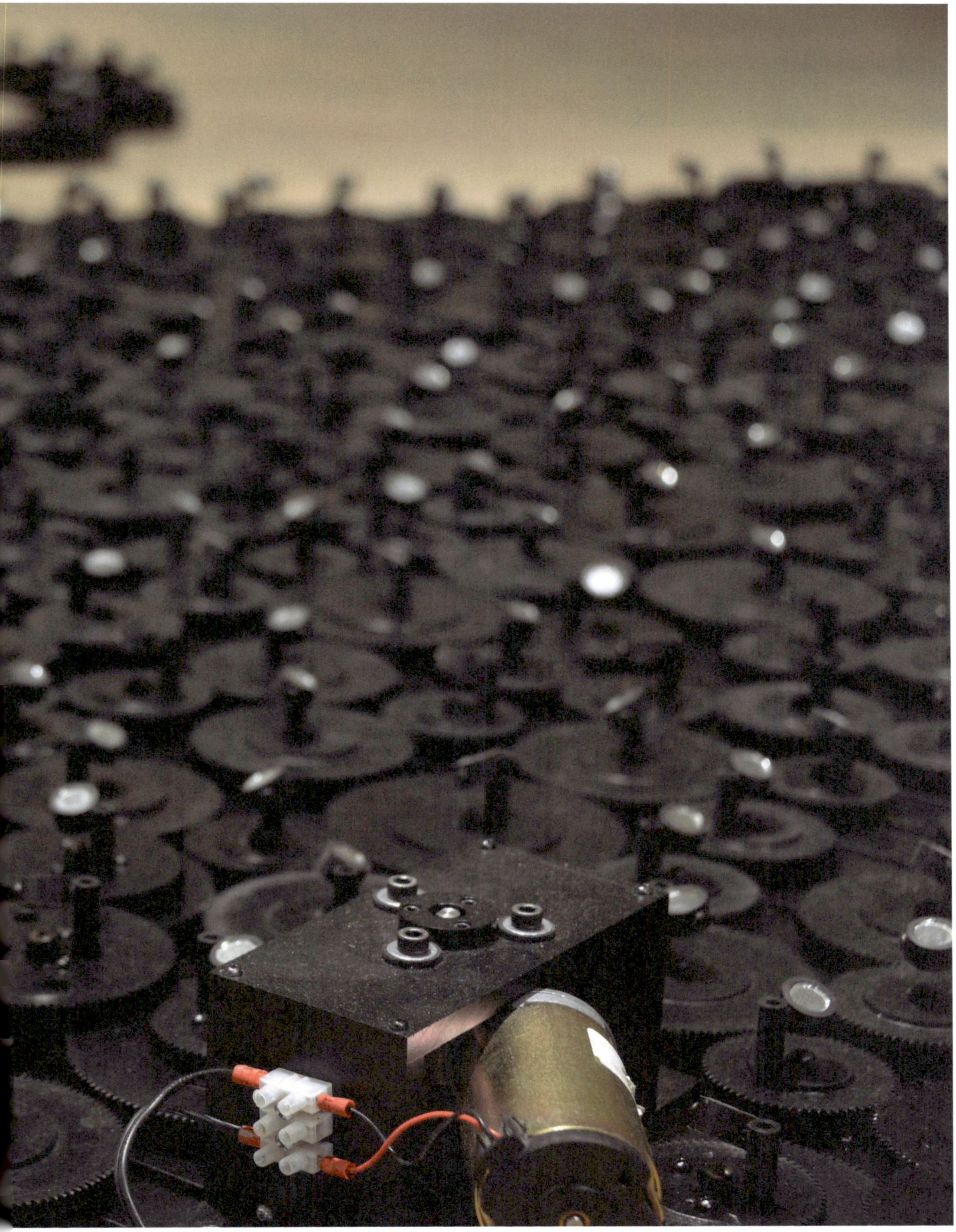

THE ANATOMY OF OIL

Marcella Durand

in a room of high round hills
and the room of water cutting
out of reach, the deep rooms
upon rooms, in bands of red,

orange, brown. We drift lower,
thirsty and surrounded by minerals,
the history of that, which:
    that, which leads us to deep
    histories, depths of hunger
    for which there is no resolve:

        As it spreads along the sills impermeable
        in best-intentioned effusion of life,
        teeming, but between intent and discovery,
        a canyon, chasm,
            that is, to explore one seam
            with intent to destroy the next.

        As bodies slowly fall, to fall
        together or one after the other.

            Like islands do, join
            in surface flux, disjointing
            the floating plates under
            us (but not written to us),
            us with notebooks in hand.

    Lines extend from continents
    to seas unseen in this chasm.
    Walls block escape and limit
    sky, but here is where
    we thirsty in the seams
    liquid with end in sight,
    are everywhere around.

It lights our lanterns
and forms our oars.
Over each other in our boat
looking for it climb
in profusion we do,
hungry as we always are
as we burn it

those lines extend from us
straight from us to the sea
and ocean floor, perfect
and simple, over what we
thought was the end of land—
giant sea-worms, tubes, mouth-
less and eyeless, under storm
of falling life, catching without
hands, into which
all bone becomes new skin

treeless in the desert for days
its beauty, inverse to its fertility
but why call it beautiful? I can't even see it

it's a national monument, enclosed within
itself and meant to be seen in a sort
of grouping, a clustering together, a kind
of huddle. Like uranium under quicksand,
it draws us into a kind of state, a hooded
state, it already knew, and within us,
cracking steam, and, on the horizon,
a signal on the top of mountains, a rope
thrown down—we leave hanging, our
hunger, the turn in the river with no exit.

thin soil that, walked on, crumbles away

***

Hoodoos and goblins, arsenic and salt springs
petroglyphs and rock houses, a trickle of river
unbordered by green—the red rock comes straight
down to the bank and is dry, as we speed by
in our 4WD wide wheelbased SUV tinted windows
following the RV with self-enclosed toilet, shower
and small venting ribbons, a TV on dashboard, camera—

but
it is even
as it is eroded
and each rain changes
the depth of the canyons
and hills melt away

and colors will not be denied
as they reach through glass and cool air.

***

One layer over the next and another laid
down over the first and others until all
is lifted up—and as such climbs higher
and creates within itself voids
and within these vast spaces
skeletons liquid and trapped
between water and air, mountain
and valley, held by such light
and substance and water, as
evaporation stops, as we
would evaporate

and within these vast spaces
underground, with two slit
windows facing frontwards
camouflaged behind hill
one room is used for
storage (we think) and the
other room is used for
experimentation (we think)
where nothing could be

held by such substance,
where evaporate where it
is not where it not is here
then where is it

***

one layer over the next and another
creates within itself voids like those
inside, those petrified objects,
heart, stomach, liver, brain, I,
combustible-engined user and allied
with you. You, fossil-fuel burner
and one of us. Us, together in this
single-stroke engined boat because
it is fast. Do you think at the end
of this river is food? Maybe we should

take this path through the woods
and see which trees filter light
and what blue heron flies off
the pond when approached,
how much we should like to meet
its eyes, how much to have
it speak, it, what is it?

and when it leaves a few feathers behind
to be later crystallized into a profit
mineralized and pumped up between it
and another. Distilled, enhanced,
combusted, then it will fit into our
projections, as convenient as could be,
is this a rant? even as everyone
laughs a little behind their hands—

a group of people scrape
cautiously at a seam, pebbles fall from the design
into a bucket—   but this structure

this eroded fractured volcanic throat and this six-sided
column fluted down the edges this peak over land of rusty
bobble-headed iron drills, thin threads poke into dry hearts—

rock-bottom cheap
it flows but
elsewhere

if gravel could be ground and rock extracted
sand crushed and siphoned, flyaway soil captured
snow and ice melted, rocky beaches compacted, and
beneath the giant thumping machines small drops well
and conglomerate into solid puddles, crushed the conveyor
belt, the ongoing oceanic drop and slide, to recycle,

this smoky place, this fire hanging above nothing, a simple
space between emittance and smokestack, clear and see-
through, a place between flame and metal, fueled and, at night,
glass shields visible from the waters, a quickly changing
reflection, and slow radiance spreads from the piers
in circles, in transforming rivets, slowly under iron.

***

when sudden flooding
drives us lower past
crushed abyss of uncertain epoch; from the rims
falls darkly staining
excess, what we
could find more
of better. Our lights
our oars, a mineral
we found in our pockets
this morning and under
the sleeping bags
even as birds cried
like a cathedral above.

O if we had an oar
and needed something
like that, and standing
in water, air takes
from us what
we thought was
soil, instead, it stains
and sits next to us

as one state sits
next to another
and topography
grows flatter and
drier and flatter
and drier.

These lines are simple, perpendicular,
contradictory, extend to the sea
in extension lined with illumination
a shell-less creature with one
foot and phosphorus—
when it falls, as it does, one after
the other, it lines a newly built
floor, and many others come after
to build a newly built floor,
which sags and is heated,
from underneath comes heat,
and is compressed and forms
something close to what we
are looking for, we look more—

what could unmoving parts say
about where this uplift came from
what caused a ship to rise in desert
we pass through, we are speeding
through, we are the middle of 120
miles without services if we run
out of gas we are in super-deep shit.
This animal, this beast, lies across
the highway and its claws leave
deep ruts our wheels bounce
and we are trapped in shimmering
light and visions of trucks and
more trucks as they approach
the undulating city of the coastal
plain, we were someplace and now
we are someplace that mega-city
to which we are speeding
120 miles and animals line
the highway and all is fenced
it is tottering toward "E"

but it
is, even
colors reach
through glass
and cool air
without
water

trapped between mountain
and valley, unvented, it lifts

What we think is oxidization the sky here feels at one with benzene and flammable
clouds noxious vapor at lunch
break clouds ignite and spread flaming drops
over school travels insatiable brilliant
appetite radioactive denizens even, as with
us, on this boat, fleeing or going toward,
there is a place we have been before

the copper river valley rises up on either side mines
shrouded by eastern forests behind the green
veil a small city and another, they are rooms
and worn down in every aspect, as every
aspect takes from us our treasure rocks
and dog-eared books. If we feel at one
with explosion, that is something
quick and sharp, and not the bobble-headed
drill in the middle of flat plains and
the fault-and-block system for miles and
distance and silence, that is not the gradual
wearing down and the thread poking
through and the tire treads on crypto-
organisms turning sand into soil. There are
warnings here to stay on the trail; there is
a warning not to go deeper.

but here we are with
tiny lanterns, in a canyon

prone to flooding, under
the bright blue lake, blue
against red rocks and the
powerboats singing and
the overpowered jetting,
such transmission, we here
under the blueness, floating
and yet going, a current
almost indecipherable, the
surface unintelligible. It
grows darker as we go,
and around us
all is blue

the substance is not secretive, but untraceable
irreducible it is found in its bases
there are bases but no pyramids
no minarets embedded within missile silos
L-amino acids and DNA dislocated from one to another
pierced by small seeping springs it has wiry leaves
taproots bright white shining purple at a certain hours
at evening, behind the coal-mining operations and over the mesas.

***

The ships in the thousands in the sea, "he
is doing the bidding of his dark oily masters,"
and the troops think of their children, or so
say the newspapers, the article written about
the woman, his wife, and what she thinks
about the skin rashes, what sort of pills
he eats with breakfast.

The ships by the thousands mass
travelling from here to there and we
in our boat flat-bottomed we wait
for a current, a breeze, "by fear
is the rule of the sea," he holds his crew
with a marble in the palm.

he says each day he gets
information about chemicals
hidden in caves here or
there, our boat floats further
between two walls, on
these walls are inscribed
the news of the day, and
sometimes far up a house
appears, it is inaccessible,
it is uninhabited, we are
not the guardians
of our own history.

They are just stones
falling down one
upon the other, there
is no "we" us here
on this boat and we
what are we but
separate one I you
another over there
no stone fits each
layer different
each other color
red and brown and
orange dark, we are
not we us no we
but then what
is oil? but millions
of creatures crushed
one into another
shoulder
to shoulder

***

rising off the river in the morning
that which can no longer be seen
and that which can no longer be said

say: there is
us in this boat and there
are you in your boats, and here
we are under you and you are
above us and you are
against our lanterns, small
circles of light fed by minerals
hungry, for minerals, and minerals
around us, above us, over us
all of us hungry and unintelligible
irreduceable but not traceable
you the same stone and us the same rock
written of us (but not to us)

our boat goes further
between two walls and
it grows darker the closer
we come to the mantle
questions: why when the
cavern-structure shifts
geyser blooms blue-orange?
why a horizon of steam evaporates
in columns and perpendicular
wind-blown from bare sand?
why did the mountain town
cover up the vapor caves with
concrete and train tracks? where
do the Utes go now? what rises off
the waters now at this time of day? we
drink everything there is to drink.

***

Is this the right itinerary? the guides split a while ago
stumbling along the sides of the crevasse,
and now we're trapped like slow-cooling
conglomerate clusters in a lava flow   a petroglyph there
where you're pointing   more houses higher than any
ladder we have   on top of each hill giving themselves a pleasant
view as plywood exudes and inhabitants
breathe essence of pulverized forest   each twig
generated by an engine   each section dark
gummy liquid now stuck to our fingers   fine webbing as
we feel along in the night for our way unguided

shoulder to shoulder we stand in our way
united and hungry our stomachs are full
we want something and take it from our neighbor
we like it big the next one should be taller
the better to see our land with all the better
to see "our" land to see our "land" the land

***

I love that rock and am amazed at how so large an uplift
can look like a ship, it is the only thing I can think of
to compare to this massive rock thing if I could only get
out of the car to take a closer look if only he would
stop and we could climb out of the car and take a closer
look if only the highway would stop and we could climb
out of the car and take a closer look if only the roads
would stop and if only we could look closer at the rock
if only we could stop the car if only we could look

***

at it closer if shoulder to
shoulder we stand closer together
but one would not be he
who has had nothing but plenty

a hero, him, who leaves faint marks
on the ground like tracks here
or through the walls lets his hammering
resound, or upstairs, he wears shoes
or checks his mailbox every night
as it swings open to reveal: deep red,
russet, dark brown water-stains, lighter
brownish-gold, bleached, some green,
silvery in contrast. Down the walls
it comes and half the landmark
with it. It's there because we want
it and if we find it we know it is
there and if it's under something
else then that something must
move or be moved and it is poison

in that it is fresh and drops slowly
without acid and small paths lead
to it as we use others to steal it.

In solubility how is water like oil
one is invisible and the other obscure
one combusts and the other is flammable
when one is put into the other it spreads
to where we stand watching how is one
like another how are you liking me how
are you where are you how are we

***

Our desert calls to your desert.
Across the earth, one desert speaks
to another, just as water wicks away
into sky. Why so fast? Our desert
is encroaching—we have one, too.
All the world should be a desert.
Because we have a desert, we deny
there is anything beneath it. If you
follow tracks into a hole, you find
a hole. Just like time goes fast on
ultimatums. We translate
rocks to find what's in
between them. Our desert is
not made of sand. Each truck
passes through a small hole.
My hand is the size of a tractor-trailer.
Between my thumb and forefinger,
a truck goes. One truck after another—
so many trucks! What are they shipping?
Dark streaks on the top
of that one, a recumbent
shiny oval. I squeeze it down
to the size of a fire ant. One truck
glows and is closely followed
by another one. Their
routes submerge and resurface
in a giant massing around a
sandy peninsula. At gas stations,
they sleep while underground

fuel tanks regenerate themselves.
My hand is the size of several
axles and slips on ice. The more axles,
the better. If I press here, my headache
goes away and fossil fuel comes out.
I buy vegetables at corner markets
and support the food wax industry.
Each apple shines like black ice. Our
apartment stinks like steam
cracking through soil. Belowground
pipes carry oil for an entire city
in the shadows of forward-leaning
ranges. Each pass harbors
a collection of movie cinemas.
The white truck glides
through traffic and belches black
smoke over people on the sidewalks.
We lean over the sides of our truck
and cast hooks at everything with
rivets in its sides. We ran out of
water a long time ago.

***

field lights piled upon each other
illuminating nothing but another field
and after that more fields until a desert
abruptly with sand ends the endless pré

it's luminous far beyond the sinking light
all that rises and casts long darkness
from either side of the cañon
for us, sun for only a few hours, for us we

need to light the path and with what
to light the walls? And how to light
the waters on all sides, this timed
sliver of day above, so quickly it turns

dark, that with darkness we counteract,
with dark substance we create a glow,

all it illuminates, past the cañon walls,
and reaching those, find another field

and in that hour which you promised
and which he promised several hours
and days ago, the high tables of land,
the generous wooded hills, the marshy

in-between lands, the sculpted flowers
hanging off the submerged cedars.
To plan certain mist, it is surprising
to exit such dunes stretching away

and find a heavy fog, hiding the road.
A small path skirting headlands,
to see such cloud of lights entering
water hissing after so many days

without water or light. Is our hunger
justified? Do dry winds spin faster
over land flattened in the last
glacial event? Will our house be lifted

and deposited in a till of glacial debris,
a giant moraine, a deep bedrock dictating
height, an island formed purely from
the garbage of others. It's the sort

of juice that increases thirst. Electrolytes
know not which way to go. It's pure
salt and a small sign obscures this arsenic
spring. Or the sky ignites from someone's

carelessly dropped corporate exemption.
Follow the dried-out shrub and you'll
find pools of sandy mineral aspiration
at which each animal gathers, excluding

ourselves: we like to have it just for us.
But, did it leave footprints in the water?
Could we find it again if we followed each
branch pointing upwards to the flammable crown?

Did this loser hang painted leaves in
each tree, thinking the birds would be misled
into building nests of bendy straws and
tinselly metal threads? Maybe a bolt

or two, but not the entire dumped car,
rusting in the front yard, or where it
rolled down the hill one night, coming
to rest against a large maple tree, just

on the edge of changing, each leaf
a reflection of the same colors displayed
when the prairie is accidentally worn away,
when one small creek grinds down to

the banded deposits below—dusty and
harboring many different kinds of
wildlife, invisible mostly to you and me,
except in that huge space where no

where it's just space, kind of, and
turrets poking into horizon, that's
the closest we could come to drinking
water and driving as quickly as we

could through juniper scent and over
ground beneath us that contains nothing,
nothing! nothing! dust and slight coloration
a bit of iron no interior lights no darkness

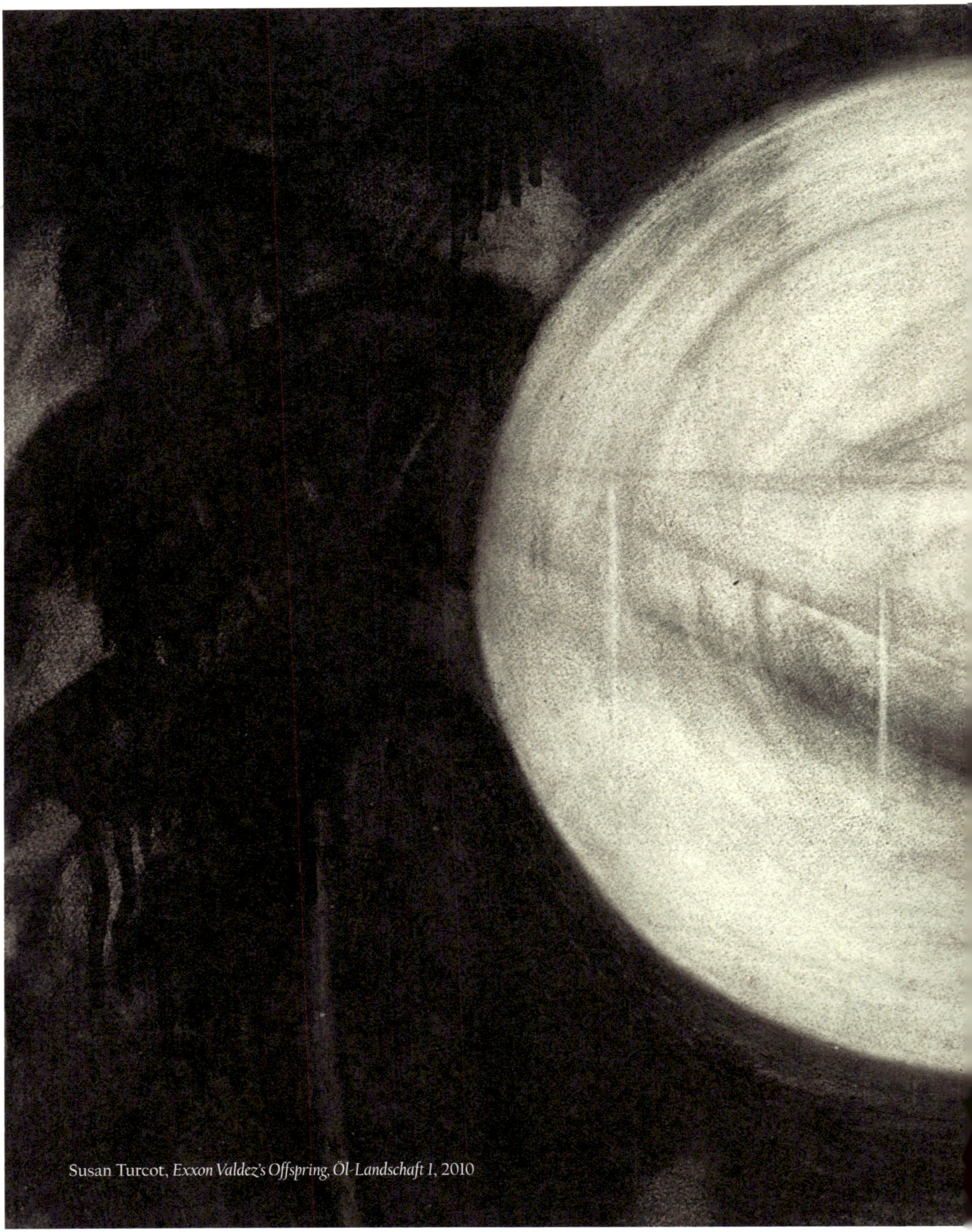

Susan Turcot, *Exxon Valdez's Offspring, Öl-Landschaft 1*, 2010

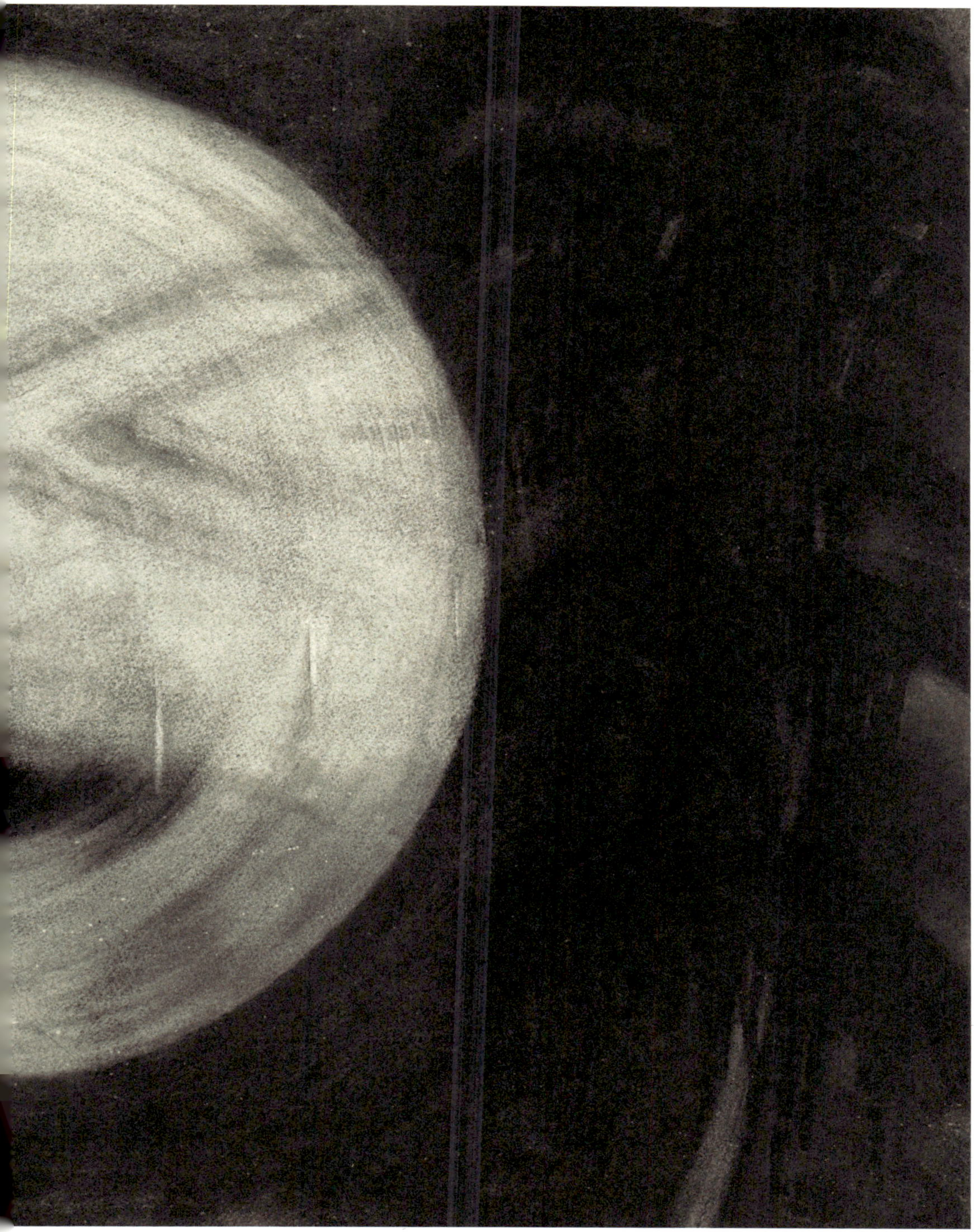

WORKS IN THE EXHIBITION

EDWARD BURTYNSKY

Alberta Oil Sands #1, Fort McMurray, Alberta, 2007

Alberta Oil Sands #2, Fort McMurray, Alberta, 2007

Alberta Oil Sands #5, Fort McMurray, Alberta, 2007

Alberta Oil Sands #6, Fort McMurray, Alberta, 2007

Alberta Oil Sands #7, Fort McMurray, Alberta, 2007

Alberta Oil Sands #10, Fort McMurray, Alberta, 2007

digital chromogenic colour prints, all works edition 1/6, 121.9 × 152.4 cm each

Courtesy Whyte Museum of the Canadian Rockies, Banff, AB

LOUISA CONRAD

The Last Arctic, I: Kendall Island Bird Sanctuary & Site of the Proposed Niglintgak Barge-Based Gas Conditioning Facility (Beginning of Pipeline), 2009
C-print, laminated and mounted on dibond, 101.6 × 101.6 cm

The Last Arctic, II: Kendall Island Bird Sanctuary & Site of the Proposed Niglintgak Barge-Based Gas Conditioning Facility (Beginning of Pipeline), 2009
C-print, laminated and mounted on dibond , 101.6 × 101.6 cm

Melting of the Mackenzie Delta from Space, June 1st 1997, 2009
graphite on paper , 55.9 × 76.2 cm

Disintegration: A Catalogue of Arctic Flowers, 2009
graphite and ink on vellum
80 drawings, approx. 12.7 × 12.7 cm each

Man on Water, 2009
video (6 min 30 sec)

Syncrude Processing Plant, Fort McMurray, Alberta, 2009
video (5 min 10 sec)

All works courtesy Anthony Greaney Gallery, Boston

ROBERT LADISLAS DERR

Oil City, 2007
video (continuous loop)

Courtesy of the Artist

JIMMIE DURHAM

Sweet light crude (Dubai–Oman), 2008
five painted barrels, 90 × 60 cm each

Courtesy Wien Lukatsch Galerie und Buchhandlung für Kunstbücher, Berlin, Germany

ROMUALD HAZOUMÈ

Dan, 1992
plastic can, telephone, rubber, 39 × 39 × 13 cm

On/Off, 1992
plastic can, seeds, metal, aluminium and copper, 52 × 43 × 24 cm

La Reine, 1995
plastic can, metal, acrylic, 25 × 17 × 19 cm

Tire Bouchon, 1996
plastic can, synthetic hair, metal, 41 × 35 × 23 cm

Ear Splitting, 1999
plastic can, brush, speakers, 42 × 22 × 16 cm

Black 2020, 2005
plastic can, copper, 42 × 22 × 16 cm

All works courtesy The Contemporary African Art Collection (C.A.A.C. — The Pigozzi Collection), Geneva, Switzerland

BRIAN JUNGEN

Dragonfly, 2008
carved gallon gasoline jug, 45.75 × 34.5 × 11.5 cm

Courtesy Reesa Greenberg, Ottawa

CAL LANE

Oil Fired Tanks, 2009
steel, 130 × 290 × 57 cm

Oil Drum Skeletal Red Map, 2011
steel, 198 × 179 × 5 cm

All works courtesy Art Mûr

DAVID LARIVIERE

Come Visit the Athabasca Tar Sands, 2012
billboard, 30.5 × 61 m

Courtesy of the Artist

ERNST LOGAR

Tillydrone, 2008

Logie, 2008

Northfield, 2008

Woodside, 2008

Torry, 2008

Aberdeen, 2008

C-prints, 156 × 125 cm each

Correspondence, 2008
A4 photocopies, exhibited on video

Scotland, Oil & Gas Pipelines, 2008
permanent marker on wall, dimensions variable

All works courtesy of the Artist

ROBYN MOODY

Power 2: Heart Lake as seen through the eyes of Manley Natland, 2009
plastic, motors, MDF, mirrors, 520 × 285 cm

Courtesy of the Artist

SUSAN TURCOT

Exxon Valdez's Offspring, Öl-Landschaft 1 – 6, 2010
charcoal and marble powder on paper, 53 × 73 cm each

Slavery I, 2010
tubes on steel/car wheels/bowls on paper, 180 × 320 × 320 cm

All works courtesy Galerie Ursula Walbröl, Dusseldorf

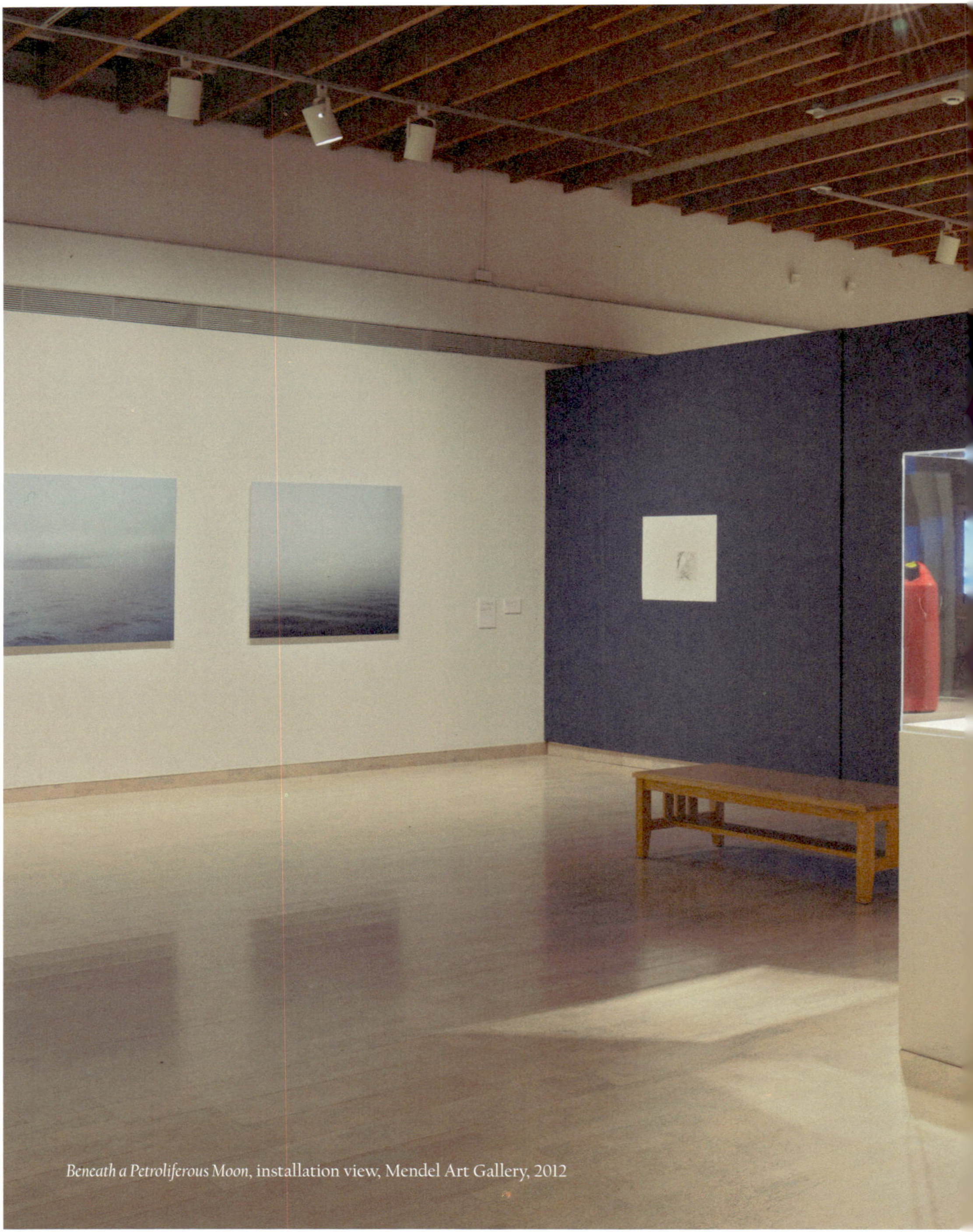

Beneath a Petroliferous Moon, installation view, Mendel Art Gallery, 2012

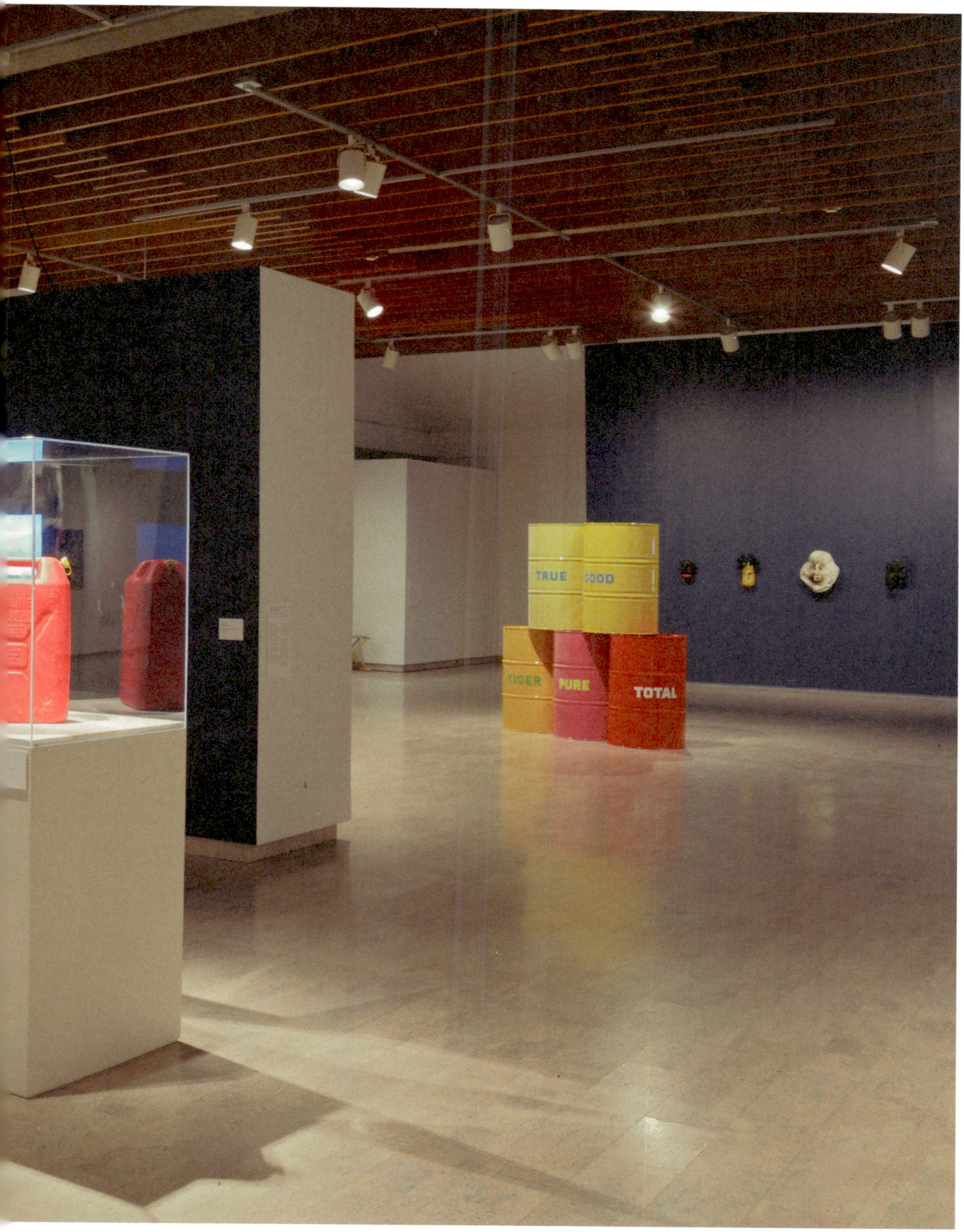
TRUE
GOOD
TIGER
PURE
TOTAL

ARTISTS

Edward Burtynsky (born in St. Catherines, ON, 1955) links his early exposure to the sites and images of the General Motors plant in his hometown to the development of his photographic work. His exhibitions include *Water* (2013) at the New Orleans Museum of Art & Contemporary Art Center (international touring exhibition), *Oil* (2009) at the Corcoran Gallery of Art in Washington, D.C. (five-year international touring show), *Manufactured Landscapes at the National Gallery of Canada* (toured 2003 to 2005), *Before the Flood* (2003), and *China* (toured 2005 to 2008). His distinctions include the TED Prize, the Outreach Award at the Rencontres d'Arles, the Flying Elephant Fellowship, Applied Arts Magazine Book Award(s), and the Roloff Beny Book Award. His prints are housed in over 50 public collections worldwide.

Louisa Conrad (born in New York City, 1982) lives in Townshend, Vermont, with her husband, the poet Lucas Farrell, a herd of sheep, one llama, three donkeys, and three Tamworth pigs at Peaked Mountain Farm, where they make cheese, cured meats, and bread. They have started their own farm, Big Picture Farm, with five goats and a line of goat milk caramels. She also has a studio in Brattlebro, VT, where she works on several projects, including a remapping of Robert Frost's poems from found ephemera. She is represented by Anthony Greaney in Boston, MA.

Robert Ladislas Derr (born in Cincinnati, 1970) uses various modes of making that centre on a barrage of questions about life and art. He makes tangible the world around him through visual manifestations that leave the viewers with a sideways glance. Presentations of his art works and performances include the Schirn Kunsthalle, Wexner Center for the Arts, and Irish Film Institute. Among his awards are the Urbana Public Arts Commission, Lower Manhattan Cultural Council, and Ohio Arts Council. Derr received his MFA from the Rhode Island School of Design and BFA from the Art Academy of Cincinnati.

Jimmie Durham (born in Arkansas, 1940) is a sculptor, essayist and poet. He became involved in the arts through the civil rights movement in the 1960s. From 1973 to 1980 he was a political organizer with the American Indian movement, serving as director of the International Indian Treaty Council and representative to the United Nations. In the early 1980s he returned to art, creating a body of work challenging colonial representations of North American Indians. In 1987 Durham moved to Cuernavaca, Mexico and in 1994 to Europe. In Europe, Durham's art has focused primarily on the relationship between architecture, monumentality and national narratives. He has exhibited in venues around the world, including the Whitney Biennial, Documenta IX, ICA London, Exit Art, and the Venice Biennale, among many others. There have been three major retrospectives: the first organized jointly by the Musée d'art contemporain, Marseilles and the Museum Voor Actuele Kunst in the Hague (2003); the second by the Musée d'art moderne de la Ville de Paris (2009); and the third by MUHKA, Antwerp (2012).

Romuald Hazoumè (born in Porto Novo, 1962) is an artist from the Republic of Benin. He is best known for his work *La Bouche du Roi*, a reworking of the 1789 image of the slave ship Brookes, made of petrol cans, audio and video elements, and spices. This work, which was widely exhibited in the UK as part of the centenary remembrance of the Slave Trade Act of 1807, is now in the collection of the British Museum. Hazoumè has exhibited his work all around the world, and is represented in The Contemporary African Art Collection of Jean Pigozzi. He lives with his family in Cotonou and maintains a studio a Porto Novo.

Brian Jungen (born in Fort St. John, 1970) has exhibited extensively in Canada and internationally, in venues including the Tate Modern (London), the New Museum (New York), the Biennale of Sydney, the Gwangju Biennial (South Korea), and Secession (Vienna). In 2006, the Vancouver Art Gallery organized a major survey exhibition of his work, which toured to New York, Montreal, Rotterdam, and Munich. His work can be found in private and public collections worldwide. He lives and works in Vancouver.

Cal Lane (born in Halifax, 1968) likes to work as a "visual devil's advocate," using contradiction as a vehicle for finding her way to an empathetic image. Having spent much of her early life as a hairdresser in her mother's Vancouver Island salon, Lane soon became interested in exploring gender roles and conceptual art. The internationally acclaimed sculptor is known for turning ordinary metal objects into lacy artworks using a blowtorch. She holds a BFA from the Nova Scotia College of Art and Design and an MFA from the State University of New York, and lives in upstate New York.

David LaRiviere (born in Edmonton, 1965) completed a Bachelor of Fine Arts degree from the University of Alberta in 1989 and an MA Fine Art degree from Goldsmiths College, University of London in 1996. Since 2008 he has worked as the Artistic Director of PAVED Arts in Saskatoon. LaRiviere's current art practice encompasses audio art, video, web art, performance and commercial intervention. His interest in a variety of media is influenced by a still-developing interest in continental philosophy. Recent solo and group exhibitions of his work have been staged in Saskatoon, Vancouver, Winnipeg, Seattle, Calgary, and Regina.

Ernst Logar (born in Klagenfurt, 1965) began his career as a photographer, before making the transition to an investigative or research-based multimedia art practice. Along with *Invisible Oil* (2009), his major projects include *Welcome to Europe — Insha'alla*, in Maribor, Spain (2012), *The End of Rememberance*, throughout Austria (2008), *Non-Public Spaces*, in Vienna, London, Yokohama, and other cities (2002-2006), and *Den Blick hinrichten*, throughout Austria (2004–2008). He lives with his family in Vienna.

Robyn Moody (born in Lethbridge, 1975) incorporates into his artworks whatever technology is required to make them work, most recently mechanics and analogue electronics. However, his work is never about the technology employed. Often humorous, beautiful, and hiding a dark secret, Moody's work explores, in whole or in part, the complex relationships between technological advances, human belief and interpretations of the world, and humanity's relationship with science, politics, and nature. Notable exhibitions in recent years include Kling and Bang (Reykjavik), Lydgalleriet (Bergen), the AND festival (Liverpool) and the 2013 Alberta Biennial of Contemporary Art.

Susan Turcot (born in Montreal, 1966) began her career working in film and installation. After studying at Middlesex University in London, UK, she lived for many years in Berlin. There, she developed her drawing practice into an intense communication of truths that lie beyond language. In all her work, including drawings, sculpture, and video, she expresses a delicate balance between destruction and stability. Turcot is represented by Arndt & Partner in Berlin and Galerie Ursula Walbröl, Düsseldorf. Her work can be found in private and public collections around the world. She resides in England.

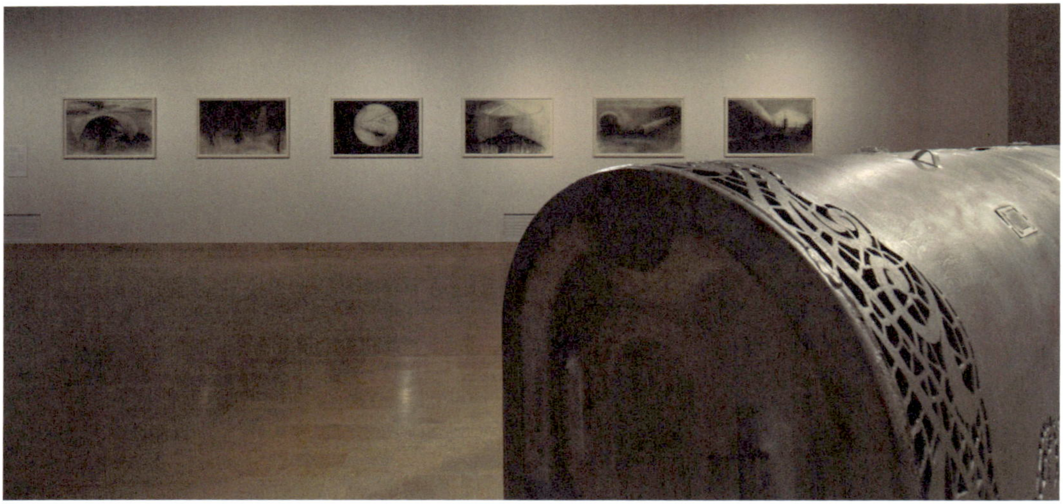

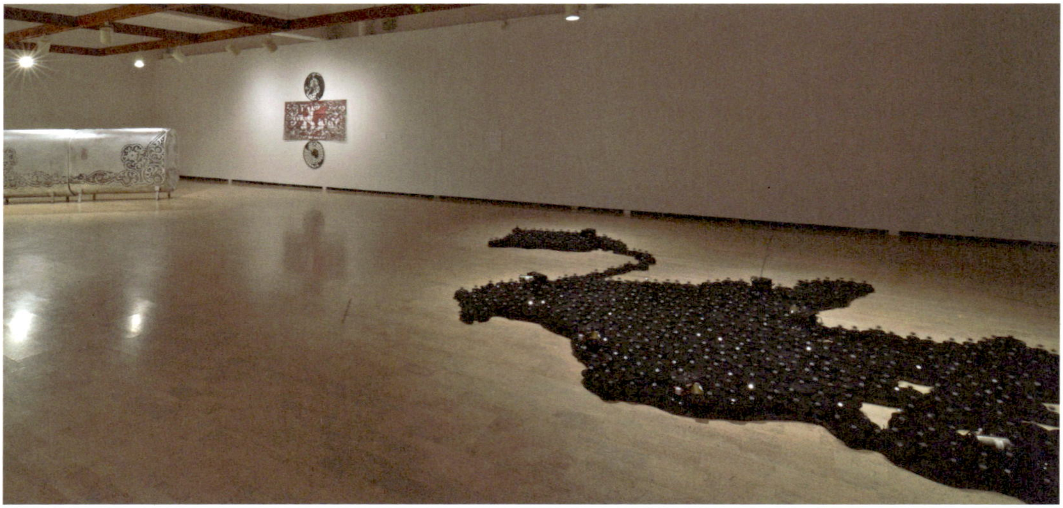

 Beneath a Petroliferous Moon, installation views, Mendel Art Gallery, 2012

CONTRIBUTORS

Jen Budney is Associate Curator at the Mendel Art Gallery, Saskatoon, where her projects have included the touring retrospectives *Jayce Salloum: history of the present* and *Ruth Cuthand: BACK TALK*, as well as *"Innocent Years": Stories and Pictures by William Kurelek, Ian W. Abdulla, and Marjane Satrapi*, and more. She writes often about epistemology, colonialism, the connections between language and land, and the uses (and abuses) of art.

Marcella Durand's most recent books are *Deep Eco Pre*, a collaboration with Tina Darragh; *AREA*; and *Traffic & Weather*, a site-specific book-length poem written during a residency at the Lower Manhattan Cultural Council in downtown Manhattan. She has recently completed a new collection, *The Prospect*, and is now writing a book-length alexandrine titled *In This World of 12 Months*. She lives in New York City with her husband, letterpress printer Rich O'Russa, and son Ismael, and works as a writer and editor for a medical nonprofit.

Pablo Neruda (born Neftali Ricordo Reyes Basoalto in 1904) was a Nobel-prize-winning poet from Chile, sometimes Chilean-diplomat, and socialist. He wrote in a variety of styles, including surrealist poems, historical epics, political manifestos, autobiography, and love sonnets. He often wrote in green ink, as a symbol of his desire and hope. He died in 1973.

Vaclav Smil is a Distinguished Professor Emeritus at the Faculty of Environment, University of Manitoba in Winnipeg. He is the author of more than 30 books on many aspects of energy, environment, food, population, risks and public policy, including *Should We Eat Meat? Evolution and the Consequences of Modern Carnivory*, *Harvesting the Biosphere: What We Have Taken from Nature*, and *Energy Myth and Realities: Bringing Science to the Energy Policy Debate*.

Editing
Trish Keegan and Richard Swain

Design
Robin Mitchell Cranfield for hundreds & thousands

Photography
All photos Troy Mamer 2012, except:
title page and pages 28, 29: Ernst Logar
pages 10 and 11, 26, 27: Edward Burtynsky, courtesy Nicholas Metivier, Toronto/Paul Kuhn, Calgary
page 18: Musée d'Art moderne de la Ville de Paris / Thierry Langro;
page 24: Robert Ladislas Derr
page 33: Casey Kaplan Gallery, New York
pages 34, 35 and pages 38, 39: Louisa Conrad
page 40: David LaRiviere
page 66: Susan Turcot / Galerie Ursula Walbröl, Düsseldorf

Printed in Canada by Hemlock, Vancouver, BC

Distributed by ABC Art Books Canada
230-372, rue Sainte-Catherine Ouest
Montreal, Quebec, Canada H3B 1A2
www.ABCartbookscanada.com

MendelArtGallery
Published by the Mendel Art Gallery
950 Spadina Crescent East, P.O. Box 569,
Saskatoon, SK, Canada S7K 3L6
www.mendel.ca

The Mendel Art Gallery is a non-profit organization supported by its members and donations and grants from the City of Saskatoon, Saskatchewan Lotteries, the Saskatchewan Arts Board, and the Canada Council for the Arts.

Library and Archives Canada Cataloguing in Publication

Budney, Jen, author
Beneath a petroliferous moon / Jen Budney ; with contributions from Marcella Durand, Pablo Neruda, Vaclav Smil.

Catalogue of an exhibition of contemporary art about the petroleum industry held at the Mendel Art Gallery in 2012.
ISBN 978-1-896359-81-6 (pbk.)

1. Petroleum in art--Exhibitions. 2. Petroleum industry and trade in art--Exhibitions. I. Durand, Marcella, writer of supplementary textual content II. Neruda, Pablo, 1904-1973, writer of supplementary textual content III. Smil, Vaclav, 1943- writer of supplementary textual content IV. Mendel Art Gallery, host institution V. Title.

N8236.P35B83 2013 704.9'4933827282 C2013-903923-6

City of Saskatoon

Saskatchewan Arts Board

Canada Council for the Arts Conseil des Arts du Canada

Saskatchewan Lotteries

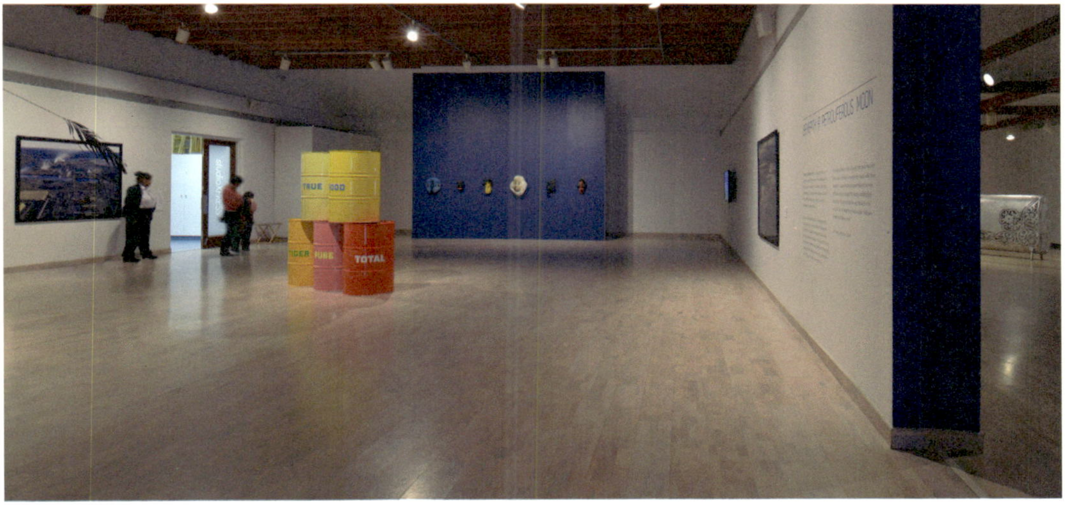

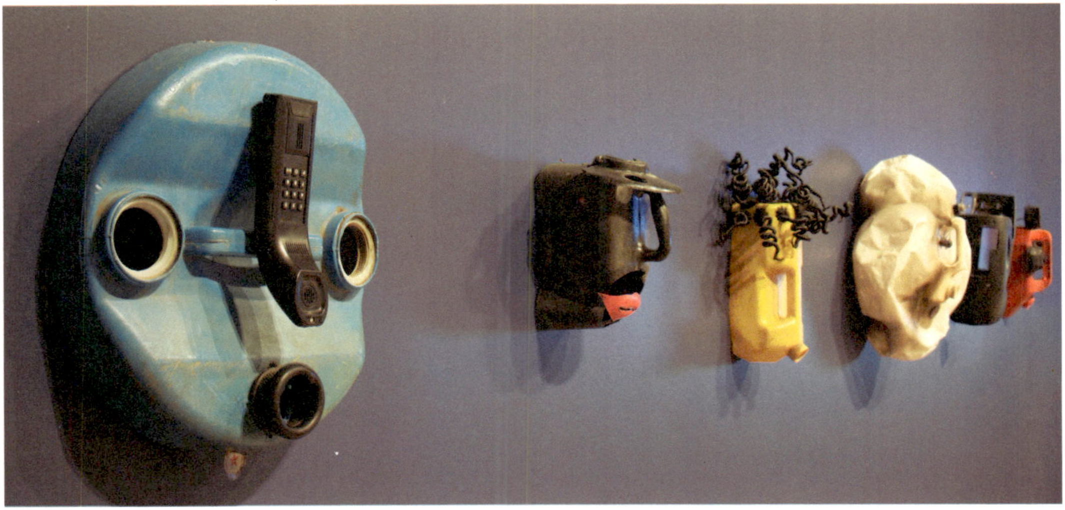

Beneath a Petroliferous Moon, installation views, Mendel Art Gallery, 2012

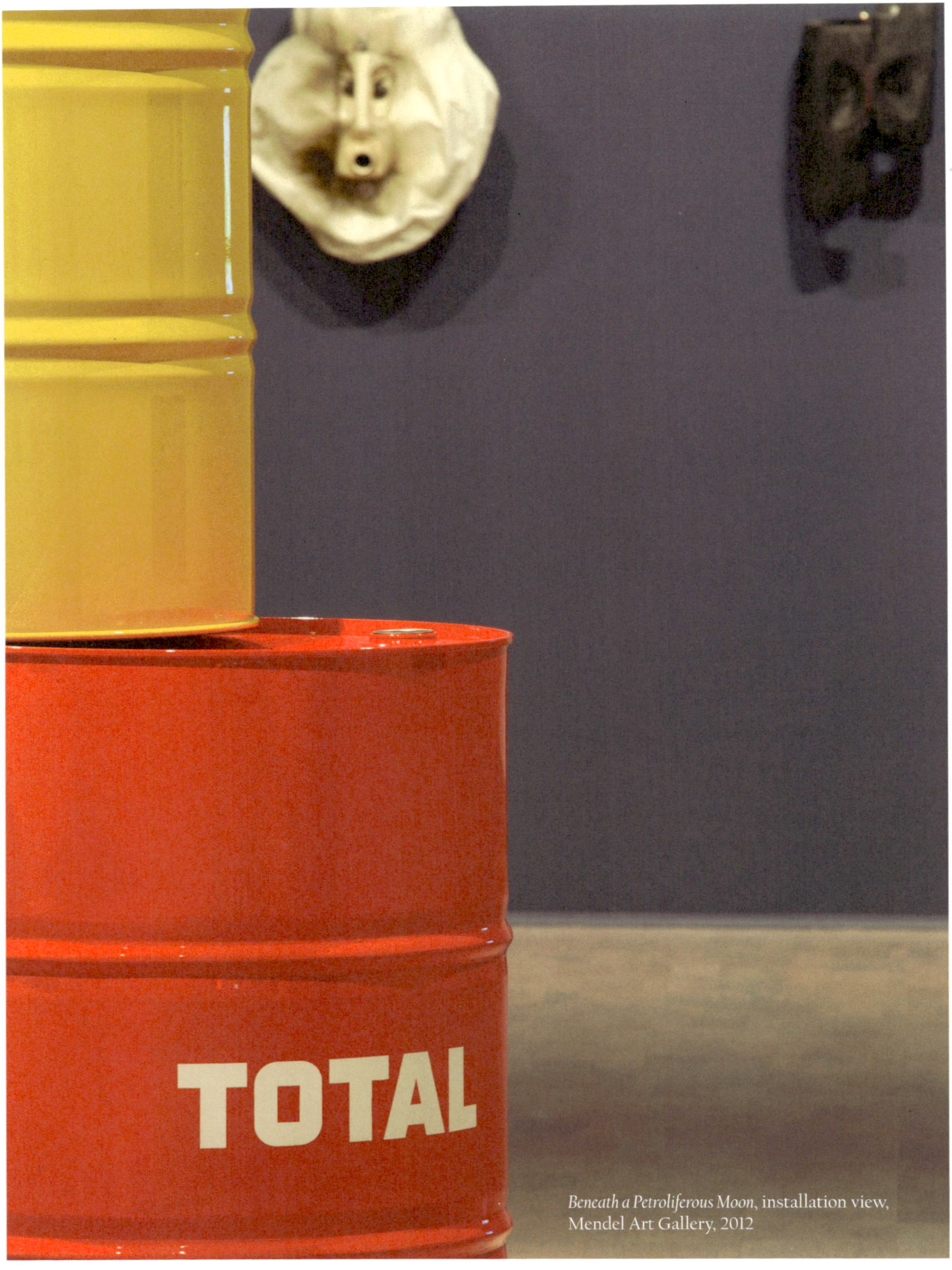

Beneath a Petroliferous Moon, installation view, Mendel Art Gallery, 2012